What Is Darwinism?

What Is Darwinism?

And Other Writings on Science and Religion

Charles Hodge

Edited and with an introduction by
Mark A. Noll & David N. Livingstone

A Division of Baker Book House Co
Grand Rapids, Michigan 49516

© 1994 by Mark A. Noll and David N. Livingstone

Published by Baker Books,
a division of Baker Book House Company
P.O. Box 6287, Grand Rapids, Michigan 49516–6287

Printed in the United States of America

All rights reserved. No part of this publication may be reproduced, stored in a retrieval system, or transmitted in any form or by any means—electronic, mechanical, photocopy, recording, or any other—without the prior written permission of the publisher. The only exception is brief quotations in printed reviews.

Library of Congress Cataloging-in-Publication Data

Hodge, Charles, 1797–1878.
    What is Darwinism? / Charles Hodge ; edited by Mark A. Noll and David N. Livingstone.
        p. cm.
    Originally published: 1874.
    Includes essays, originally published 1862–1863, and excerpts from Systematic theology, originally published 1872–1873.
    Includes bibliographical references and index.
    ISBN 0-8010-6792-8
    1. Natural selection—Philosophy. 2. Evolution (Biology)—Religious aspects—Christianity. 3. Evolution (Biology)—Philosophy. I. Noll, Mark A., 1946– . II. Livingstone, David N., 1953– . III. Hodge, Charles, 1797–1878. Systematic Theology. Selections. IV. Title.
QH375.H64   1994
575.01′62—dc20   94-9426

To
Jim Moore

Contents

Acknowledgments *9*

Introduction: Charles Hodge and the Definition of "Darwinism" *11*

Three Brief Notices *49*

    Hodge's First Published Comment on Darwin (1862) *49*

    Response to the *New York Observer* (1863) *51*

    Excerpts from the *Systematic Theology* (1872–73) *56*

What Is Darwinism? *61*

    Theories as to the Nature of the Universe *64*

    Mr. Darwin's Theory *77*

    The Distinctive Element of Darwinism: Rejection of Teleology *89*

    Relation of Darwinism to Religion *129*

Asa Gray's Review of *What Is Darwinism?* *159*

Bibliography *171*

Index *175*

Acknowledgments

The editors would like to thank Allan Fisher of Baker Books for his initial interest in this project and also to Ray Wiersma for superb copyediting. Without several industrious student assistants—Greg Clark, Andrew Graham, Jon Farrar, Jason Mitchell, Lori Willemsen, and especially Anne Van Kuiken—this book could never have been completed. For help with translations, we thank Arthur Rupprecht and Sara Miles. The book's dedication reflects a long history of specific encouragement on this project and more general stimulation to hard thinking on the questions which also engaged the mind of Charles Hodge.

Introduction: Charles Hodge and the Definition of "Darwinism"

This edition of Charles Hodge's writings on science, featuring *What Is Darwinism?* is being published in conjunction with a forthcoming companion volume entitled *B. B. Warfield on Scripture and Evolution*. New editions of older works are always arguments as well as exhibitions, and these two are no exception. They will serve a historical purpose by making readily available two significant statements that a bygone era produced on the subject of religion and science. Just as important, however, these two books are intended as illustrations of a type of Christian thinking that is, regrettably, less common than it once was.

At the end of the twentieth century, when controversies over evolution excite even more passion than they did when Hodge and Warfield were alive, the specific answers they proposed for reconciling modern scientific research and historic Christianity may still commend themselves to some. Even more, these writings—with their patient analysis and unhesitating confidence in both science and Scripture—offer modern Christians a better way than extremist, anti-intellectual, or paranoid combat against the scientific establishment. At the same time they offer scientific despisers of traditional biblical faith consequential examples of a responsible respect for science that arises directly out of Christian belief itself.

Charles Hodge (1797–1878) and Benjamin Warfield (1851–1921) were the two most influential theologians at Princeton Theological Seminary during the first century of that institution's existence. During their tenure Princeton was widely regarded as the nation's center of Presbyterian intellectual life, and this at a time when, as Eugene Genovese has recently commented, "the power of the Presbyterian divines . . . would be hard to exaggerate."[1] Hodge taught at Princeton from 1822 until his death; Warfield, who studied under Hodge, succeeded Hodge's son as professor of theology in 1887 and remained at Princeton until he died in 1921.[2]

The Princeton Approach

Hodge and Warfield reached contrasting conclusions on questions of cosmology: Hodge thought that Charles Darwin's formulation of evolution was tantamount to atheism; Warfield held that a comprehensive formulation of evolution could be reconciled with an orthodox Christian faith based on an inerrant Bible. Yet these differing conclusions are not as important as the goal, the mentality, the tradition, the substantial intellectual convictions, and the view on the role of theology in American civilization that the two shared.

The common goal was to preserve the harmony of truth. Hodge and Warfield refused to countenance any permanent antagonism between two of the fundamental realms of knowledge: what humans, by God's grace, can discover about the natural world (which owes its origin to God); and what they can learn, again by grace, about the character and acts of God from special revelation in the Bible.[3]

---

1. Eugene D. Genovese, *The Slaveholders' Dilemma* (Columbia: University of South Carolina Press, 1992), 1.

2. For general orientation on Princeton in the era of Hodge and Warfield, see John O. Nelson, "The Rise of the Princeton Theology: A Genetic History of Presbyterianism until 1850," Ph.D. diss., Yale University, 1935; Thomas J. Wertenbaker, *Princeton 1746–1896* (Princeton: Princeton University Press, 1946); Deryl F. Johnson, "The Attitudes of the Princeton Theologians toward Darwinism and Evolution from 1859 to 1929," Ph.D. diss., University of Iowa, 1969; W. Andrew Hoffecker, *Piety and the Princeton Theologians: Archibald Alexander, Charles Hodge, and Benjamin Warfield* (Grand Rapids: Baker, 1981); Mark A. Noll, ed., *The Princeton Theology, 1812–1921: Scripture, Science, and Theological Method from Archibald Alexander to Benjamin Warfield* (Grand Rapids: Baker, 1983); and David F. Wells, ed., *The Princeton Theology* (Grand Rapids: Baker, 1989).

3. In their belief in the compatibility of natural and revealed knowledge, Hodge and Warfield were maintaining a long-standing Christian tradition affirming the "two books" by which God has made himself known. This approach to religion and science had been explicitly promoted by Francis Bacon, and so had substantial currency in America, where Bacon's name was revered. See James R. Moore, "Geologists and Interpreters of Genesis in the Nineteenth Century," in *God and Nature: Historical Essays on the Encounter between Christianity and Science*, ed. David C. Lindberg and Ronald L. Numbers (Berkeley: University of California Press, 1986), 322–35.

Introduction: Charles Hodge and the Definition of "Darwinism"

Their common mentality was that of scholars. Hodge and Warfield were alike committed to thorough reasoning. They thought it was a Christian duty to use their minds fully to understand the world. They did not set reasoning about the physical world and interpretations of divine revelation in opposition; rather, they held that properly qualified assertions of the human intellect and properly understood conclusions from Scripture are complementary. As a consequence, they were patient in unpacking detailed arguments in theology as well as philosophy, and they abhorred merely rhetorical responses to complicated intellectual problems.

The tradition they shared was the long-standing enterprise of Christian scholarship. It is appropriate that E. Harris Harbison's noteworthy book *The Christian Scholar in the Age of the Reformation* was first presented as Stone Lectures at Princeton Theological Seminary, for Harbison's insightful analysis describes almost exactly what Hodge and Warfield had attempted at that seminary in earlier generations. As Harbison puts it, "The calling of a Christian scholar . . . may be to shoulder any of . . . three major tasks . . . or some combination of the three: (1) to restudy the Hebraic-Christian tradition itself, (2) to relate this tradition to the surrounding secular culture and its tradition . . . , and (3) to reconcile faith and science, in the broadest sense of the word."[4] Hodge and Warfield were engaged in each of these tasks to at least some extent. Both were students of the Scriptures, and Warfield especially was a student of the students of Scripture. Both were eager to address the moral and religious questions their contemporaries were asking. And both were greatly concerned about reconciling the study of nature and the study of revelation "in the broadest sense of the word."

Hodge and Warfield also held common intellectual convictions. Theologically, they were Calvinists who maintained traditional Reformed convictions about most subjects, including nature. That is, they held that the world owes both its origin and its ongoing operation to the direct activity of God. They believed that God is responsible for the orderliness of natural processes, that the human ability to discern this order in nature is a gift from God, and that investigations of nature testify to the work of a purposeful designer. They also

---

4. E. Harris Harbison, *The Christian Scholar in the Age of the Reformation* (New York: Scribner, 1956), 5.

felt that Scripture provides reliable general information about the physical world.

Philosophically, the Princeton theologians were committed to the principles of commonsense reasoning as these had been imported to North America in the eighteenth century by Scotsmen like John Witherspoon (president of Princeton College from 1768 to 1794) and then developed by a host of American commentators. Against the skepticism of philosophers like David Hume, commonsense philosophy featured trust in ordinary human intuitions. Proponents of this philosophy drew on sophisticated arguments by Scots like Thomas Reid and popularizations thereof in works like the *Encyclopaedia Britannica* as edited by Reid's follower Dugald Stewart. With such support, Americans easily turned aside doubts about the reality of the self and of basic cause-and-effect connections. Especially in America, the philosophy of common sense was joined with a trust in scientific induction (the Baconian method) as a way of adjudicating issues in philosophy and theology as well as the study of nature. (Hodge and Warfield shared at least some of the American enthusiasm for Isaac Newton, as the doyen of modern science, and Francis Bacon, as the most famous early promoter of an epistemology of induction.)[5] In the American context, reasoning on the basis of commonsense principles occupied much of the place that reliance on tradition had held for the churches of Europe. The Princetonians' Calvinistic convictions about the debilitating character of sinfulness did not always fit smoothly with their commonsense philosophy. Yet that philosophy was common intellectual coinage in nineteenth-century America, and they were among the American intellectuals who put it most skillfully to use.[6]

In many ways, the Princetonians' theology of nature and their philosophy of common sense were typical of most American Christians during the nineteenth and twentieth centuries. But unlike at

---

5. Especially illuminating on such matters are Henry F. May, *The Enlightenment in America* (New York: Oxford University Press, 1976); and Theodore D. Bozeman, *Protestants in an Age of Science: The Baconian Ideal and Antebellum American Religious Thought* (Chapel Hill: University of North Carolina Press, 1977).

6. Hodge's debt to Baconian common sense surfaces nowhere more clearly than in the introduction to his *Systematic Theology*, 3 vols. (New York: Scribner, 1872–73), 1:10, 13: "The Bible is to the theologian what nature is to the man of science. . . . In theology as in natural science, principles are derived from facts, and not impressed upon them."

least some other conservatives, they also held that Scripture does not need to be interpreted literally when it refers to nature. Perhaps even more untypically, they held that the findings of science should be enlisted to help discover proper interpretations of Scripture.

On questions of public life, Hodge and Warfield shared a belief that a properly articulated theology is essential for the flourishing of society. As upholders of American republican ideals, they felt that *virtue* in citizens is prerequisite for the health of a society. And virtue of a reliable and long-lasting sort, in turn, can be assured only by the presence of active Christian faith. Christian faith, in turn, requires both intellectual and spiritual grounding; it is a matter of properly reasoning about the truth alongside properly acting upon the truth. The authority accorded to theologians like Hodge and Warfield, as well as their own sense of value to the community at large, depended on their ability to promote ideas and practices that encouraged virtue. Because they reasoned this way, the Princetonians naturally held that the faithful promotion of sound theology is critical, not just to the church narrowly conceived, but to the future of society as a whole. With such views, questions about the place of science (and scientists) were much more than academic questions. Wrong views on the nature of the world, as well as inappropriate deference to authorities who do not value Christian traditions, would strike at the heart of civilization as well as of the gospel.[7]

Alike as they were on many matters, Hodge and Warfield did differ on one specific question: Could the type of evolution described by Charles Darwin in *The Origin of Species* (first edition, 1859) and *The Descent of Man* (first edition, 1871) be construed as a manifestation of the orderly design with which God had made the world? Hodge said no, Warfield yes.

---

7. The convictions of Hodge, Warfield, and their colleagues deserve to be taken seriously by modern students, even as modern students realize the Princetonians' own stake in their arguments. Along with similarly placed intellectuals at Andover and Yale in New England, these theologians enjoyed a privileged position in American social life and were regarded as authorities on ethical, intellectual, and religious issues. It answers no questions about religious or scientific truth to note that the status of these men depended upon their ability to convince others that they spoke the truth. Nonetheless, modern students, in their own search for the truth, owe it to themselves to examine not only what these theologians said, but also how their hegemony in the social order depended on convincing their peers that *they* spoke the truth.

The works of Charles Hodge that are included in this volume highlight not just his conclusions about Darwin's theories, but the larger constructions of his thought on religion and science. These writings, especially *What Is Darwinism?* will repay close attention for several reasons. Most obviously, they are intrinsically significant as key documents from the pen of one of the nineteenth century's most influential theologians. They address, moreover, questions that were, and remain, among the most pressing intellectual issues of the modern era. Further, they illuminate broader issues at stake in discussions of Darwin and Darwinism. By tracing critical reaction to *What Is Darwinism?* at the time of its publication and in later debates over Darwin's ideas, we confront the immensely complicated question of just what "Darwinism" means and the extraordinary general significance of those debates for the relationship between Christianity and modern science. Before we can explore those matters, however, it is necessary to offer a brief outline of Hodge's career and lifelong involvement with science, including some consideration of how science fit into the framework of his general convictions.

The Life and Science of Charles Hodge

Charles Hodge, born on December 27, 1797, was the son of a Philadelphia physician and merchant who had served with distinction in the American War for Independence.[8] His mother, Mary Blanchard Hodge, was descended from New England patriots. Both parents were serious-minded Presbyterians who wanted to bring up Charles and his older brother, Hugh Lenox Hodge, to honor learning, patriotism, and the Westminster Confession of Faith. When Hodge was less than a year old, his father died from an infection contracted in one of the periodic yellow-fever epidemics that beset Philadelphia after the Revolutionary War. Mary Blanchard Hodge then kept a boarding house in Philadelphia, receiving assistance from relatives and—until the economic downturn preceding the War of 1812 destroyed its value—living off a small legacy from her

---

8. A modern critical biography of Charles Hodge is a pressing desideratum, especially given the recent work that has been done at Princeton Theological Seminary by archivist William Harris to catalogue Hodge's extensive papers. In the interim the biography by his son is still the fullest account available: Archibald Alexander Hodge, *The Life of Charles Hodge* (New York: Scribner, 1880).

husband. Zeal for the education of Hugh and Charles led her in 1812 to Princeton, New Jersey, where she enrolled them in the College of New Jersey (later Princeton University).

At Princeton, Charles Hodge was instructed by the college's president, Ashbel Green, and another clergyman, Archibald Alexander, who was also new to Princeton, having arrived in 1812 as the first professor at the Presbyterians' new theological seminary.[9] Green, who had been Hodge's pastor in Philadelphia, was attempting to promote the balance between learning and piety that he had learned under Princeton's great Revolutionary Era president, John Witherspoon. Alexander, a native of Virginia who had been an itinerant preacher and then a pastor in Virginia and Philadelphia, took a special interest in Hodge, especially after the latter entered the seminary in 1816. From both Green and Alexander, Hodge learned a combination of biblical fidelity, seventeenth-century Calvinist theology, intense practical piety, and commonsense reasoning that would mark his own long career.

Hodge graduated from the seminary in September 1819, and returned the next year to teach the biblical languages. In May 1822 the Presbyterian General Assembly appointed him to a regular position, a post that he held until his death fifty-six years later. Through his long career at Princeton Seminary, Hodge's theological determination, tireless pen, breadth of interests, and favorable social connections made him not only one of the century's most influential theologians, but a major public influence as well. When he celebrated his fiftieth anniversary as a Princeton professor in 1872, Hodge personally had instructed more theological students than had attended any other theological institution in the United States.

The influence of Hodge's writings was even greater than that of his teaching. In 1825 he founded the journal that has been popularly known as the *Princeton Review*. As its indefatigable editor and principal author, Hodge organized a committed band of colleagues who with him used its pages to defend the confessional Presbyterian faith they cherished as the clearest expression of the gospel. Hodge's other writings included commentaries on several books of the New Testament, several works on Presbyterian ecclesiastical affairs, and

---

9. On this era at Princeton see Mark A. Noll, *Princeton and the Republic, 1768–1822: The Search for a Christian Enlightenment in the Era of Samuel Stanhope Smith* (Princeton: Princeton University Press, 1989).

numerous expositions for lay people, of which *The Way of Life* (1841) was the most notable for its limpid prose and clear exposition. In the early 1870s Hodge summed up a lifetime of classroom instruction with his three-volume *Systematic Theology*, a work that dominated Presbyterian theological education until the 1930s.[10]

As a theologian, Hodge's point of view was consistent. He contended for a Calvinism that had been defined in the sixteenth and seventeenth centuries. He proclaimed the dangers of unchecked religious experience, whether in the form of sophisticated European Romanticism or rough-hewn revivalism of the American frontier. The heart of Hodge's interest was the Augustinian picture of human salvation that he felt had been most fully stated by the Protestant Reformers. The defense of this theology inspired his weightiest polemics. What most troubled him were positions that undercut high Calvinistic convictions about divine sovereignty in salvation or that valued too highly the moral capacities of human nature.

Hodge's life was filled with practical services to Presbyterians and other Protestants. The last of these services was particularly symbolic for our purposes. One month before he died on June 19, 1878, Hodge attended church for the last time, having been called upon to pray at the funeral in Washington, D.C., of his longtime friend, Joseph Henry, founder of the Smithsonian Institution.[11]

Hodge's lifelong interest in science was partly a familial legacy. Although he had no personal recollection of his father's medical practice, his own knowledge of medicine was considerable. In the interim between graduation from seminary and his return to Princeton as an instructor, Hodge studied Hebrew for one year in Philadelphia. But also in that period he attended lectures in anatomy and physiology at the University of Pennsylvania.[12] Hodge's closest

---

10. John M. Mulder and Lee A. Wyatt, "The Predicament of Pluralism: The Study of Theology in Presbyterian Seminaries since the 1920s," in *The Pluralistic Vision: Presbyterians and Mainstream Protestant Education and Leadership*, ed. Milton J. Coalter, John M. Mulder, and Louis B. Weeks (Louisville: Westminster/John Knox, 1992), 41. For an indication of the continuing influence of Hodge's *Systematic Theology*, see Rick Nutt, "The Tie That No Longer Binds: The Origins of the Presbyterian Church in America," in *The Confessional Mosaic: Presbyterians and Twentieth-Century Theology*, ed. Milton J. Coalter, John M. Mulder, and Louis B. Weeks (Louisville: Westminster/John Knox, 1990), 251–54.

11. A. A. Hodge, *Life*, 579–80.

12. Ibid., 68.

friend and a constant correspondent for more than fifty years was his brother, Hugh Lenox Hodge (1796–1873), who became a notable gynecologist, professor of obstetrics at the University of Pennsylvania, and author of a substantial text on obstetrics.[13] Not surprisingly, then, Charles retained an interest in human physiology, therapeutics, and medical research throughout his life, not least because of his own suffering from severe rheumatic symptoms during early adulthood.[14]

Hodge's interest in science, as well as his commitment to the harmony of religion and science, was promoted by Ashbel Green and Archibald Alexander, the major educational influences of his formative years. Green was extremely proud of the scientific achievements of his sons, one of whom became a notable chemist.[15] His inaugural address as president of the College of New Jersey, "The Promotion of Science in Union with Piety," spoke for the larger purposes of his tenure at the college.[16] For his part, Alexander had been tutored by William Graham, a graduate of Princeton who ardently promoted Witherspoon's advocacy of the Newtonian system as the appropriate method for resolving all intellectual problems. Alexander himself taught science during a stint as president of Hampden-Sydney College in Virginia, regularly read scientific materials, and during his time at Princeton Seminary often attended discussions coordinated by science instructors at the College of New Jersey.[17]

Hodge's interest in scientific matters had one practical consequence that especially impressed his friend Joseph Henry. This was his practice, carried on from about 1830 to his death, of recording the daily temperature, wind direction, and cloud cover in Princeton. These daily observations constituted, according to Henry, the

---

13. See Harold Speert, "Hugh Lenox Hodge and His Vaginal Pessary," in *Obstetrics and Gynecology in America: A History* (Chicago: American College of Obstetricians and Gynecologists, 1980), 225–27. The editors thank Dr. James H. Maxwell for this reference.

14. During his one extended absence from Princeton (1826–28), when Hodge undertook theological study on the Continent, he made a point of visiting scientific as well as theological experts (A. A. Hodge, *Life*, 190–91).

15. See Edgar Fahs Smith, "Jacob Green Chemist, 1790–1841," *Journal of Chemical Education* 20 (1943): 418–27.

16. On Green's intellectual commitments see Noll, *Princeton and the Republic*, 180–82, 211–12, 272–88.

17. See especially Bozeman, *Protestants in an Age of Science*, 40–41.

only persistent meteorological record in Princeton, at least during Henry's time there.[18]

Hodge, in fact, sustained a particularly warm friendship with Henry, notable pioneer in magnetism and electricity, founder of the Smithsonian Institution, longtime professor at the College of New Jersey (1832–48), sometime trustee of Princeton Theological Seminary (1844–51), and a serious Presbyterian layman. Hodge recruited Henry to write for the *Princeton Review*,[19] included Henry in the informal meetings where the editorial business of the *Review* was carried out, worked hard to keep Henry at the College of New Jersey when other institutions tried to lure him away, and took special delight when Henry was introduced to the Presbyterian General Assembly in 1843 as a man who carried out scientific duties with a due sense of piety.[20] Another indication of the relationship the two enjoyed is that when Hodge's rheumatic leg gave him the most difficulty, he applied electricity to the afflicted limb from a machine that Henry had invented. Henry was not altogether pleased with this experiment, which in fact did not relieve Hodge's condition.[21] The closeness of their relationship was also indicated by their ability to transcend differences of opinion. Henry, after initially doubting the possibility of compatibility between any form of evolution and traditional Christianity, eventually came to a reluctant acceptance of a Christianized form of evolution.[22] Even though Hodge could not agree, he remained on very cordial terms with his scientific friend.

In addition to his avocational scientific interests, Hodge, at least from the late 1840s, regularly lectured and wrote on issues concerning the relationship of science and Scripture.[23] In 1859 he exploited the publication of a book by J. L. Cabell of the University of Virginia on the unity of humanity to publish on the subject. In a lengthy essay

---

18. A. A. Hodge, *Life*, 236; and *The Papers of Joseph Henry*, ed. Nathan Reingold et al., 6 vols. to date (Washington: Smithsonian Institution Press, 1972–), 5:267.

19. Henry published three papers in the *Review*: "The British Scientific Association" (1841); "The Coast Survey" (1845); and "Observations on Colour Blindness" (also 1845). For Hodge's appreciation of Henry see "Joseph Henry," *Biblical Repertory and Princeton Review: Index Volume from 1825 to 1868*, 194–200.

20. A. A. Hodge, *Life*, 239; and *Papers of Joseph Henry*, 2:426; 5:42, 159 n. 4, 264–65, 353.

21. *Papers of Joseph Henry*, 2:90n, 240–42, 266–67.

22. See Ronald L. Numbers, *The Creationists* (New York: Knopf, 1992), 11.

23. For example, "The Mosaic Account of Creation," January 1849 lecture, Charles Hodge Papers, Speer Library, Princeton Theological Seminary.

Hodge contended (as he would also in his March 1863 letter to the editor of the *New York Observer*; see pp. 53–56) that "the church is willing to meet men of science on equal terms." The church on several occasions had wisely changed its interpretation of the Bible in response to new scientific discoveries. But if the church should defer to science, so also should scientists take account of "moral and religious truths." The sum of the matter, as he saw it, combined nicely Hodge's twin confidence in an infallible Scripture and intuitions of human consciousness: "The grand objection after all to any theory of diversity of species or of origin among men, is that all such theories are so opposed to the authority of the Bible, and to the facts of our mental, moral, and spiritual nature."[24] In light of Hodge's later arguments concerning Darwin, it is ironic that in this essay he spent much of his time attacking Louis Agassiz for defending polygenism, the view that humanity was made up of multiple species (and a view that was regularly used to denigrate nonwhite races).[25] Later Hodge would enlist Agassiz's assistance in refuting Darwin, for Agassiz became the best-known American opponent of evolution by natural selection.

Hodge's writings on scientific themes mostly promoted various strategies of harmonization that had developed over the preceding two centuries. That is, he tried to show how biblical accounts (when reinterpreted in light of modern scientific discoveries) had anticipated the conclusions of the scientists. As might be expected in light of his debt to commonsense forms of reasoning, Hodge also made much of what human consciousness can reveal about the nature of our existence in the world.[26]

In sum, Charles Hodge was a respectable amateur on scientific questions. Throughout his life he read fairly widely on such matters, including European works in French and German. He was particularly alert to the theological implications of scientific issues, espe-

---

24. Charles Hodge, "The Unity of Mankind," *Biblical Repertory and Princeton Review* 31 (June 1859): 106, 105, 148.

25. Ibid., 105–6, 135.

26. For example, Charles Hodge, "The Nature of Man," *Biblical Repertory and Princeton Review* 37 (Jan. 1865): 111: "The idea of substance is one of the primary truths of the reason. It is given in the consciousness of every man, and is therefore a part of the universal faith of men." Hodge also touches on scientific matters in "Examination of Some Reasonings against the Unity of Humankind," *Biblical Repertory and Princeton Review* 34 (July 1862): 435–64.

cially when such matters were taken up in American intellectual circles. Hodge never presented himself as a scientific expert. Nor did he display the same penetrating intellect on the philosophy of science as, for example, Jonathan Edwards had displayed in the preceding century. But, proper qualifications having been made, Hodge's scientific understanding was fairly broad and his expertise fairly deep on questions at the intersection of science and theology.

Charles Hodge's Stake in Science

The Princeton theologians, beginning with Alexander and continuing through Hodge to Warfield, were interested in science for several interrelated reasons. First was confessional. As Calvinists they believed the physical world was an arena in which God manifested his power and glory. Scientific research was a way of finding out more about the world God had made, but also about the God who had made the world. Second was apologetic. The Princeton theologians knew that in the wake of Newton and the mechanical philosophy, science was being increasingly used to attack traditional Christian faith. If others used science to discredit the Christian faith (as Thomas Paine did when Archibald Alexander and Ashbel Green were young pastors, and as Darwinians seemed to be doing in Hodge's day), it was the responsibility of mature Christian thinkers to show the error of such abuses. Third was social and ideological. Christian appropriation of science was critical for the health of civilization in America. If science (or any other false claimant to be the source of ultimate value) undercut faith in God, evil would inevitably proliferate, public virtue would retreat, and civilization would be imperiled.

The Princetonians' views of the relationships of science, theology, and the civil society also implied much about their own role. Alexander, Hodge, Warfield, and their colleagues were remarkably pious people; personal testimonies abound to their unusual humility. At the same time they also possessed an extraordinarily lofty conception of their own vocation. They were guardians not just of theology, and not only of relationships between science and theology, but of truth and of civilization. Part of their concern about the spread of sub- or anti-Christian uses of science was, thus, concern about themselves. If scientists with no concern for the traditions that the theologians defended would succeed in taking their place as public arbiters of the culture's most pressing questions, it was obvious that they would also be displaced from their positions of cultural authority.[27]

Science in general, therefore, was important to Charles Hodge both because of what he believed and because of who he was. Especially as the pace of scientific discovery quickened in the nineteenth century, and as alternatives to Christian appropriations of scientific knowledge grew more forceful, the stakes grew rapidly. Without this background it is not possible to understand why Charles Hodge responded to Charles Darwin as he did.

How *What Is Darwinism?* Came to Be Written

Despite Hodge's stake in the controversy, *What Is Darwinism?* was not written until nearly a decade and a half after the publication of *The Origin of Species* in 1859. Hodge took early note of the *Origin* in a footnote to an 1862 essay on the unity of humankind (see pp. 50–51). And Hodge treated Darwin at some length in his *Systematic Theology* (1872–73). Why then did Hodge pick up the subject again in 1874?

The immediate precipitating agent was the sixth international meeting of the Evangelical Alliance, held in New York City on October 2–12, 1873, where Hodge spoke out on the nature of Darwinism. A local situation may have moved him to this step. The Scottish philosopher and theologian James McCosh had come from the Queen's College in Belfast in 1868 to become president of the College of New Jersey. McCosh was as concerned as Hodge, who was then the senior member of the college board of trustees, about the possible damage to Christianity from the new evolutionary views. Unlike Hodge, however, McCosh was able to reconcile evolution and traditional Christian faith. In particular, McCosh contended for an evolution in which the inheritance of acquired characteristics (that is, Lamarckianism) guides organic progress along paths that God has preordained. As he put it, "I have been defending Evolu-

---

27. For this process in America see Alexandra Oleson and John Voss, eds., *The Organization of Knowledge in Modern America, 1860–1920* (Baltimore: Johns Hopkins University Press, 1979); and Dorothy Ross, *The Origins of American Social Science* (New York: Cambridge University Press, 1991). For Britain see Frank M. Turner, "The Victorian Conflict between Science and Religion: A Professional Dimension," *Isis* 69 (1978): 356–76; and for issues involving evolution, Adrian Desmond and James R. Moore, *Darwin* (New York: Warner, 1991).

tion, but, in so doing, have given the proper account of it as the method of God's procedure, and find that when so understood it is in no way inconsistent with Scripture."[28]

Some historians have suggested that Hodge spoke out at the Evangelical Alliance and then in *What Is Darwinism?* to attack McCosh's position.[29] A better interpretation, however, in light of the facts that Hodge had welcomed McCosh warmly in 1868, that he never spoke out publicly against McCosh, and that the two retained a cordial relationship, is that McCosh's preoccupation with the issue of reconciling Christianity and modern scientific conclusions was a stimulus for Hodge to advance his own thinking on Darwin and Darwinism.[30]

Hodge's participation in the meetings of the Alliance deserves full attention, not only as the lead-in to his book on Darwinism, but also to show the broad impact of public Protestantism at this time in the United States. The Evangelical Alliance had been formed in 1847 to draw together Protestant churches and individuals from Europe, Britain, and North America.[31] By 1873 it included also a solid representation of missionaries and nationals from what is now called the Third World. When the Alliance met in New York City, it was headlined in local papers like the *New York Times* and in general was treated with the kind of consideration that at the end of the

---

28. *The Life of James McCosh: A Record Chiefly Autobiographical*, ed. William M. Sloane (New York: Scribner, 1897), 234.

29. For example, Frederick Gregory, "The Impact of Darwinian Evolution on Protestant Theology in the Nineteenth Century," in *God and Nature*, ed. Lindberg and Numbers, 375; and Herbert Hovenkamp, *Science and Religion in America, 1800–1860* (Philadelphia: University of Pennsylvania Press, 1978), 211–14.

30. The compatibility between McCosh and Hodge is well argued in Bradley J. Gundlach, "The Evolution Question at Princeton, 1845–1888," M.A. thesis, Trinity Evangelical Divinity School, 1989. It is noteworthy that McCosh was asked to read the Scripture at Hodge's funeral in 1878 (A. A. Hodge, *Life*, 584). Just as Hodge would later claim that Asa Gray was not a Darwinian (because Gray continued to find teleology in Darwin's natural selection), so too he may have reasoned about McCosh (and more persuasively so, since McCosh favored the Neo-Lamarckian view of the inheritance of acquired characteristics).

31. For a good general history see Philip D. Jordan, *The Evangelical Alliance for the United States of America, 1847–1900: Ecumenism, Identity and the Religion of the Republic* (Lewiston, N.Y.: Edwin Mellen, 1982). *History, Essays, Orations, and Other Documents of the Sixth General Conference of the Evangelical Alliance Held in New York, October 2–12, 1873*, ed. Philip Schaff and S. Irenaeus Prime (New York: Harper, 1874), provides full documentation for the New York meeting.

twentieth century is reserved for political conventions.[32] After the gathering in New York, foreign guests traveled in a body to Washington, D.C., where they held a brief service at the Capitol and were greeted by President Grant at the White House.

As it happened, the Alliance met at a critical juncture in the history of Darwin's ideas in the United States. Historian Jon Roberts has recently shown that the early 1870s marked a point of transition in American attitudes toward Darwin.[33] During the first decade after the publication of Darwin's *Origin,* most American scientists rejected its conclusions as poorly grounded science. It was thus no great problem for America's theologians to provide convincing arguments of their own against the book's conclusions. By the 1870s, however, the scientific tide had turned. As more and more scientists accepted evolution of some sort (though not necessarily Darwin's natural selection), theologians had to make a critical choice. They had to decide whether to follow their standard practice since the day of Newton by adjusting Christian doctrine, in this case to evolution, as earlier they had done in response to cutting-edge science concerning the age of the earth and the nebular hypothesis. Or they had to reverse their practice of reconciling Scripture and the best modern science and attack the new biological cosmology. At stake were the issues themselves, but also the social relevance of the theologians. In an earlier America they had earned the right to speak out on public issues, as well as on theology, by showing how science (the age's great, and supposedly neutral, form of reasoning) could support traditional Christian teaching. The early 1870s constituted a day of decision. Would the alliance of science and theology hold, or would it break? The implications for science, for theology, and for the place of theologians in American society were great.

Charles Hodge's main purpose in traveling to New York for the conference was to offer one of the keynote addresses. That speech, "The Unity of the Church Based on Personal Union with Christ," raised the eyebrows of some ardent denominationalists who felt that Hodge had treated the distinctives of their own bodies too lightly.[34]

---

32. The *Times,* for example, regularly devoted five or six columns of small print to the Alliance each day of its meeting in New York.
33. Jon H. Roberts, *Darwin and the Divine in America: Protestant Intellectuals and Organic Evolution, 1859–1900* (Madison: University of Wisconsin Press, 1988).
34. Jordan, *Evangelical Alliance,* 93–96; for Hodge's speech see *Sixth General Conference,* ed. Schaff and Prime, 139–44.

But in general it was well received.

Hodge remained for other sessions, one of which sparked his public comment on evolution. On Monday, October 6, he attended the session of the philosophical section at St. Paul's Episcopal Church, where James McCosh presented the first paper, "Religious Aspects of the Doctrine of Development." In this address McCosh restated his conviction that a distinctly biblical slant could be put on at least some modern theories describing the earth and its inhabitants as the result of evolution, and he opined that, instead of simply denouncing the theory of evolution, "religious philosophers might be more profitably employed in showing . . . the religious aspects of the doctrine of development."[35]

In the discussion that followed, the first comment was from the Reverend George W. Weldon of London, who, while opposing Darwin, maintained that the Bible is not meant to teach details of science. Then the Reverend J. C. Brown of Berwick-upon-Tweed in England arose, identified himself as a botanist, and declared that he found the developmental hypothesis very helpful in accounting for natural phenomena.

This was the cue, it seemed, for which Hodge had been waiting. Rising amid what the *Times* called "an immense concourse of people,"[36] Hodge addressed to Brown a question that may really have been intended for McCosh:

> I don't stand here to make any speech at all. I rise simply to ask Dr. Brown one question. I want him to tell us what development is. [At this point the *Times* recorded "Applause."] That has not been done. The great question which divides theists from atheists—Christians from unbelievers—is this: Is development an intellectual process guided by God, or is it a blind process of unintelligible, unconscious force, which knows no end and adopts no means? In other words, is God the author of all we see, the creator of all the beauty and grandeur of this world, or is unintelligible force, gravity, electricity, and such like? This is a vital question, sir. We can not stand here and hear men talk about development, without telling us what development is.[37]

---

35. *Sixth General Conference,* ed. Schaff and Prime, 270.
36. For coverage of the day's meetings see *New York Times,* 7 October 1873.
37. *Sixth General Conference,* ed. Schaff and Prime, 318.

Introduction: Charles Hodge and the Definition of "Darwinism"

The crux of this impromptu intervention would become the crux of *What Is Darwinism?* How should "development," or "Darwinism" as Hodge later put it, be defined?[38] Brown responded to Hodge by quoting the Westminster Catechism's statement that God is Lord over everything. And the *Times* recorded applause for his answer as it had for Hodge's question.

Subsequent sessions featured some of the brightest lights of American science. Arnold Guyot, a renowned geologist from the College of New Jersey who also lectured to the Princeton seminarians, was well known as an opponent of Darwin's natural selection. His paper—"Cosmogony and the Bible; or, The Biblical Account of Creation in the Light of Modern Science"—tried to harmonize the account in Genesis 1 with geological developments. When the Rev. Alexander Burnett of Aberdeen asked Guyot directly whether the days of creation were geological periods, Guyot affirmed that they were.

At that point another important Christian scientist was called to the floor. John William Dawson of McGill University, Montreal, was a well-known geologist and dedicated Presbyterian layman.[39] Dawson had won Hodge's respect for both his diligent fieldwork on the geology of North America and his forthright Christian profession. Eventually the two became close friends. Shortly after the Alliance conference Hodge exerted considerable effort in trying to recruit Dawson for the College of New Jersey faculty, since in Hodge's mind Dawson was the best anti-Darwinian scientist in North America.[40]

---

38. Even as this debate was taking place, the label *development* was gradually being abandoned by self-professed Darwinians (like Darwin himself) who objected to the progressivist implications of the word. Darwin felt he was describing organic change, not necessarily progress, development, or the ascent of civilization.

39. Ronald Numbers has noted that during the 1870s Guyot and Dawson were just about the only practicing scientists of note in North America to oppose organic evolution (*Creationists*, 7–17). But both of these scientists eventually moved in the direction of developmentalism. See especially John F. Cornell, "From Creation to Evolution: Sir John William Dawson and the Idea of Design in the Nineteenth Century," *Journal of the History of Biology* 16 (1983): 137–70.

40. Dawson visited Hodge in Princeton on December 16 and 20, 1874 ("Memoranda, Aug. 1, 1866–March 31, 1878," 169, Charles Hodge Papers, Speer Library, Princeton Theological Seminary). In a letter dated April 16, 1878 (Hodge Papers, Princeton University Library), Dawson spelled out to Hodge his reasons for remaining in Montreal (mainly, to aid Protestants in largely Catholic Quebec).

Dawson entered the discussion on October 6 when someone asked him directly "whether there is any necessary antagonism between the Darwinian system and the Christian religion." Dawson waffled, but did say that on the factual level Darwin had not yet shown how new species came about. At that juncture Hodge arose to assist his friend. Again his point hung on the matter of definition:

> My idea of Darwinism is that it teaches all the forms of vegetable and animal life, including man and all the organs of the human body, are the result of unintelligent, undesignating forces; and that the human eye was formed by mere unconscious action. Now, according to my idea, that is a denial of what the Bible teaches, of what reason teaches, and of what the conscience of any human being teaches; for it is impossible for any such organ as the eye to be formed by blind forces. It excludes God; it excludes intelligence from everything. Am I right?

So led, Dawson responded cautiously: "I think Darwin would not admit so much as has been said, and yet I believe his doctrine logically leads to that conclusion." Conjecturing as to what Darwin meant by natural selection, Dawson concluded, "So stated, the doctrine is *not a result of scientific induction*, but a mere *hypothesis*, to account for facts, not otherwise explicable except by the doctrine of creation."[41] In light of how clearly Hodge had focused the issue on matters of definition, it is noteworthy that the session ended with a speech to the effect that anyone attempting to discriminate theistic from atheistic evolution should be cautious in defining terms.

Hodge's interest had clearly been aroused. He returned to Princeton the next day, October 7. He then seems to have gone right to work on an expansion of his impromptu remarks, an exercise that was interrupted on October 13 when approximately 275 members of the Evangelical Alliance stopped off for a visit in Princeton on their way to Washington. By January 14, 1874, Hodge was able to read "a paper on Darwinism" to an informal meeting of his colleagues in Princeton.[42] And within two or three months the book was out.

41. *Sixth General Conference*, ed. Schaff and Prime, 319–21.
42. "Memoranda," 144–45, 151.

The Basic Concerns of *What Is Darwinism?*

Charles Hodge's conclusion to *What Is Darwinism?* is justly renowned for its characteristic absence of ambiguity: "We have thus arrived at the answer to our question, What is Darwinism? It is Atheism" (p. 156). In *What Is Darwinism?* Hodge is making a very specific point. He is attacking not evolution as such, not even the principle of "natural selection" that Darwin had proposed in *The Origin of Species*, but what Hodge considered to be the "ateleological" character of Darwin's conception of natural selection.

These terms are critical. "Natural selection" is the process whereby variations that make it more likely that individual members of a given species of plant or animal will survive are passed on to succeeding generations, while those variations that make it less likely that individual members will survive tend to die out. "Ateleology" is the absence of evident design or purpose in the physical world. (The presence of design or purpose is, by contrast, "teleology.")

Hodge, let it be said, had his doubts about both evolution as such and natural selection as a principle of biological descent. But he also knew that other theological conservatives of his generation, like James McCosh, were making their peace with these principles. (A. A. Hodge and B. B. Warfield, his successors in Princeton's chair of theology, would also eventually conclude that these principles can fit within a biblically conservative interpretation of nature.)[43] Whatever doubts Hodge himself felt about evolution or natural selection, however, these matters were not his main concern when he wrote the book on Darwinism.

Hodge's concern, rather, was the question of design or purpose in nature. In response to the proposal that natural selection works without teleology, Hodge was adamant. Darwinism must be regarded as atheism because it is an account of the history of nature

---

43. A. A. Hodge's biography of his father is terse almost to the point of silence on the subject of Charles Hodge's anti-Darwinian statements. There is one sentence (p. 549) on the debates at the Evangelical Alliance conference as well as one noncommittal sentence (p. 576) on the book: "In 1874 he published a small book entitled 'Darwinism,' in opposition to the prevailing doctrine of Atheistic Evolutionism." A. A. Hodge later reviewed favorably one of Asa Gray's last efforts to show how Darwinism could be teleological, and so implicitly took Gray's side against his father (A. A. Hodge, review of *Natural Science and Religion*, by Asa Gray, *Presbyterian Review* 1 [1880]: 586–89).

that rules out divine design. Darwin uses the word *natural,* Hodge observes, to mean "antithetical to supernatural," and thus "in using the expression *Natural Selection,* Mr. Darwin intends to exclude design or final causes" (p. 85). For Hodge this is the heart of the matter. Here the very essence of Darwin's theory lies exposed. That "this natural selection is without design, being conducted by unintelligent physical causes," Hodge declares, is "by far the most important and only distinctive element of his theory" (p. 89). The denial of design is the very "life and soul of his system" (p. 121). And again: "It is however neither evolution nor natural selection which gives Darwinism its peculiar character and importance. It is that Darwin rejects all teleology or the doctrine of final causes. . . . [And] it is this feature of his system which brings it into conflict not only with Christianity, but with the fundamental principles of natural religion" (p. 92). In sum, any scientific conclusion that even seems to go beyond description of natural processes to a denial of God's superintendence of the cosmos is not only contradictory to Scripture and inconceivable by human common sense, but disreputable science as well. In the belief that the argument of Darwin's *Origin of Species* crossed that line, Hodge wrote his book.[44]

At the same time, *What Is Darwinism?* communicates a great deal about Hodge's general attitude toward science. As a strict empiricist and a devotee of induction, he maintained the greatest respect for scientific research. Regularly he differentiated such research from cosmological speculations or metaphysical hypotheses that were, by their nature, more suspect than the careful results of patient observation. Given this dedication to empirical science, it is not surprising that Hodge praised Darwin for the careful character of his actual research: "He is simply a naturalist, a careful and laborious observer,

---

44. Even this brief sketch shows why Hodge's work, though it concludes that "Darwinism . . . is Atheism," is nonetheless not welcomed by modern "creationists." Hodge, in point of fact, was far less militant on the more general questions of evolution than his famous quotation suggests. Therefore, it comes as no surprise that *What Is Darwinism?* is not cited in the works of modern "creationists" like Henry M. Morris, John C. Whitcomb, and Duane Gish, although they regularly supply lengthy lists of other authorities supporting their arguments. See Henry M. Morris, *The Troubled Waters of Evolution* (San Diego: Creation-Life, 1974); idem, *History of Modern Creationism* (San Diego: Master, 1984); idem, *The Biblical Basis for Modern Science* (Grand Rapids: Baker, 1984); idem and John C. Whitcomb, *The Genesis Flood: The Biblical Record and Its Scientific Implications* (Philadelphia: Presbyterian and Reformed, 1961); Duane Gish, *Evolution: The Challenge of the Fossil Record* (El Cajon, Calif.: Creation-Life, 1985).

Introduction: Charles Hodge and the Definition of "Darwinism"

skillful in his descriptions, and singularly candid in dealing with the difficulties in the way of his peculiar doctrine" (p. 78). Nor, in light of Hodge's respect for what he considered the scientific aspects of Darwin's work, is it as surprising as it might otherwise seem that Hodge in the last weeks of his life, as his grandson tells us, reread Darwin's *Voyage of a Naturalist* and commented, "That is a very remarkable and delightful book."[45]

Again, given this confidence in the inductive fruit of empirical research, it is not surprising that Hodge believed it is occasionally necessary for the church to change its mind on how it interprets the Bible with respect to nature. Such changes of mind, however difficult and sometimes traumatic they might be, are periodically necessary because the findings of scientists deserve respect. If something can be reasonably established as a scientific fact, it should be accepted as a truth given by God, since God is the source of all factuality. A well-established scientific truth, in turn, might legitimately influence interpretations of Scripture and even show that traditional biblical interpretations are mistaken. Hodge, in other words, held steadfastly to the truth of the Bible, but not to the inerrancy of those who interpret the Bible. Without fear or favor he expressed the conviction that scientists make mistakes, but also that the solidly grounded results of their labors should be considered truth from God with which interpretation of the Bible must be reconciled.

Hodge's commitment to the results of scientific research explains why he devoted a substantial section of *What Is Darwinism?* (pp. 138–49) to dealing with relevant facts. This is the section where Hodge enlists authorities against both natural selection and evolution. The most striking thing about this section, however, is its open-ended character. Once having established that a Christian view of the world must be teleological, Hodge presents his best interpretation of what he considers to be facts. But he is far less dogmatic at this point than in the earlier discussion of design. The tone suggests that, although he is not yet convinced, he considers evolution and natural selection to be debatable questions.

Finally, Hodge's training and disposition also allowed him to enlist the givens of his own mind—"intuitions which are infallible,

---

45. William Berryman Scott, *Some Memories of a Palaeontologist* (Princeton: Princeton University Press, 1939), 75. The title in Scott's account seems to be a conflation or confusion of two of Darwin's books, *Journal of a Naturalist* and *Voyage of the Beagle*.

laws of belief . . . the testimony of consciousness and . . . the intuitions of the reason and conscience" (p. 137)—as a different kind of fact alongside empirical observations. What Hodge took to be empirical facts, in other words, had much to do with how he had learned to look at human nature and the physical world.

The Reception of *What Is Darwinism?*

Reviews of *What Is Darwinism?* appeared more rapidly than is now the case for serious books. Already by May 1874, less than three months after publication, Asa Gray brought out his long rebuttal to Hodge (pp. 160–69), and several other notices were appearing as well. It would be incorrect to suggest that *What Is Darwinism?* created a sensation. In fact, the book has received more serious attention from historians since World War II than it did from Hodge's contemporaries in the period of its publication. Yet theological journals did immediately recognize it as a consequential argument.

Representatives from a wide range of churches hailed the book as a conclusive answer to Darwinism. As might be expected, the hometown journal praised Hodge for answering the question of his title "beyond all doubt" (*Presbyterian Quarterly and Princeton Review* 3 [July 1874]: 558–59). The *Baptist Quarterly* (8 [1874]: 374–75) also thought that "Dr. Hodge's fresh and lucid treatise" had successfully sustained its thesis. The Episcopalian *American Church Review* (26 [1874]: 316–19) praised the book for telling "clearly, and not censoriously, what Darwinism is *now*," even as the reviewer held out hope that a formulation of evolution might emerge that would be more compatible with Christianity. The *Methodist Quarterly Review* (56 [July 1874]: 514–16) averred that "Dr. Hodge brings overwhelming proofs" to demonstrate "that evolution as by [Darwin] taught negatives theology and involves Atheism." The same reviewer, however, could not resist a subtle dig: he asked why Hodge when attacking Darwin insisted so much on the power of human intuition, while, on the other hand, intuition so solidly opposed Hodge's Calvinistic doctrines on free will and predestination. The longest favorable review showed what strange bedfellows the struggle against Darwin could make. It came from L. J. Livermore in the *Unitarian Review and Religious Magazine* (3 [March 1875]: 237–50), where a view of design even more positive than Hodge's was combined with a full exposition of the progressive, ameliorative ascent of life on the earth.

The review that would have been most interesting was, unfortunately, never written. In a letter to Charles Darwin on June 16, 1874, Asa Gray enclosed a copy of his review that had recently appeared in *The Nation*. A remark Gray made is pertinent to our own consideration of the problems in defining "Darwinism": "You will see what uphill work I have in making a theist of you, 'of good and respectable standing.'"[46] Darwin's response was noncommittal: "I read with interest your semi-theological review, & have got the book, but I think your review will satisfy me. The more I reflect on this subject, the more perplexed I grow."[47] Unlike many later commentators, Darwin seems to have realized that the definitional question Hodge raised was no simple matter.

Of contemporary reviews Gray's was the most important, and so we have published it in full. Gray was a key figure in the nineteenth-century discussion of religion and evolution. From his position as a distinguished Harvard botanist, Gray commanded the respect of scientists. From his profession as a trinitarian Congregationalist who affirmed the ancient Christian creeds, Gray won the trust of many believers. That Gray was among the first American champions of Darwin's ideas and established a close working relationship with the British naturalist added to the importance of his opinions. For our purposes, what is most interesting is that Gray considered himself a Christian Darwinian who succeeded in showing how Darwin's formulation of evolution had disclosed a previously unrecognized form of natural teleology.

Two points on which Gray insists in his review merit careful attention: (1) Hodge was really more concerned about defending divine intervention in nature than about simply defending teleology; (2) Darwin's views were teleological. Both assertions were matters that later commentators highlighted in their treatment of *What Is Darwinism?* Both point directly to what Hodge considered critical—how should Darwinism be defined? That question is far more complicated than first appears.

---

46. *Letters of Asa Gray,* ed. Jane Loring Gray, 2 vols. (Boston: Houghton Mifflin, 1893), 2:646.

47. Darwin to Gray, 30 June 1874. Used by permission of *Archives, Gray Herbarium,* Harvard University, Cambridge, Massachusetts, USA.

The Dilemmas of a Definition

Hodge's project in *What Is Darwinism?* can appropriately be considered an extended exercise in definition. His purpose, very largely, was to provide a precise definition of Darwinism in order to determine its nature and thereby to ascertain just what the lineaments of an appropriate Christian response might be.

Before turning to Hodge's own definition of Darwinism, however, it is worth pausing to reflect on what precisely is at stake in the formulation of definitions. Definitions are by their nature boundary-marking or boundary-making exercises. They are designed to demarcate the true from the false, the legitimate from the illegitimate, the relevant from the irrelevant. Accordingly, the control of definitions is of enormous consequence for intellectual debate, since definitions position both ideas and people on particular sides of debates. For Hodge this enterprise was of crucial significance, for on the definition of Darwinism hung issues of great moment. It did much more than just determine who was, and who was not, a legitimate scientist; it helped draw the line between the faithful and the infidel.

As we have noted, Hodge was very clear in his mind about what Darwinism was: it was atheism. By thus defining Darwinism as a purely naturalistic account of origins, Hodge set the terms of the debate and was thus in a position to adjudicate who was a Darwinian and who was not. This certainly led to interesting results, and none more so than the case of Asa Gray. Gray, in his own eyes, was a teleological Darwinian. So how did Hodge deal with him? Hodge applied the categories of his definition: "America's great botanist, Dr. Asa Gray, avows himself an evolutionist, but he is not a Darwinian" (p. 155). To Hodge, Gray was either mistaken or just plain mixed-up if he considered himself a Christian Darwinian. That label simply had no meaning. The term *Christian evolutionist,* by contrast, was legitimate. As Hodge insisted: "There are Christian men who believe in the evolution of one kind of plants and animals out of earlier and simpler forms; but they believe that everything was designed by God, and that it is due to his purpose and power that all the forms of vegetable and animal life are what they are. But this is not the question. What Darwin and the advocates of his theory deny is all design" (p. 101).

With Gray and other Christian evolutionists like James McCosh, Hodge still had his differences. His understanding of recent scientific research did not give him the confidence that Gray as a scientist and McCosh as a philosopher-theologian displayed in regard to the facts of the matter. As Hodge read the Bible and as he took in what naturalists said about the development of species over time, he thought there were problems with the evidence. He could not see, as he put it, how "the theory of evolution can be reconciled with the declarations of the Scriptures." But, perhaps with reference to Gray or even more likely to McCosh, he conceded that "others may see it and be able to reconcile their allegiance to science with their allegiance to the Bible" (p. 138).

Present-Day Ambiguity

What should we make of Hodge's effort? Did he have a correct sense of the essence of Darwinism? Was Darwin's theory inherently antipathetic to design and thus to orthodox Christianity? Opinions on this issue are certainly divided, even today. Consider as evidence the fact that twentieth-century judgments on Hodge's book, which almost always reflect the way in which the critics themselves define Darwinism, have been strikingly divided. With two different kinds of commentators, Hodge's effort has appeared weak or even risible. With two other types, Hodge's work is much more estimable.

1. The most dismissive accounts of *What Is Darwinism?* can themselves be readily discounted. Despite Hodge's persistent and extended effort in *What Is Darwinism?* to focus attention on the question of teleology, a set of critics has insisted that Hodge was merely a hidebound biblical literalist with a mind deformed by inbred conservatism. As one of them put it, Hodge "measured Darwin's hypothesis by its conformity to Genesis, literally interpreted."[48] Such views simply miss the point.

---

48. William E. Phipps, "Asa Gray's Theology of Nature," *American Presbyterians* 66 (Fall 1988): 171. Similar peremptory dismissals of Hodge's argument—again without seeming to have understood the central point about teleology writ large in the book—are found in Richard Hofstadter, *Social Darwinism in American Thought*, rev. ed. (Boston: Beacon, 1955), 26; Hovenkamp, *Science and Religion in America*, 212, 216; J. David Hoeveler, Jr., *James McCosh and the Scottish Intellectual Tradition* (Princeton: Princeton University Press, 1981), 274; and R. Jackson Wilson, ed., *Darwinism and the American Intellectual: An Anthology*, 2d ed. (Belmont, Calif.: Wadsworth, 1989), 32. A few other accounts of *What Is Darwinism?* though insightful and charitable toward

2. A second class, which also has taken Hodge's effort lightly, consists of commentators who are convinced *a priori* that Darwin's proposals were self-evidently teleological. In the view of those who hold to this conviction, Hodge's effort to demonstrate the ateleological character of Darwinism was doomed to the failure it deserved. One of these critics notes correctly that "Hodge refused to take seriously the vague references to a Creator in *Origin*," but then goes on to upbraid Hodge for concluding that Darwinism was tantamount to atheism.[49]

3. By contrast, several modern historians of science regard Hodge's *What Is Darwinism?* as one of the most perceptive works of its period *precisely because* it defines Darwinism as an ateleological cosmology. One of these historians, Neal Gillespie, regards the nineteenth-century debate over Darwinism as part of a much broader argument. As he sees it, Hodge was correct: a conflict did in fact exist between the older creationist accounts of natural phenomena and what Gillespie calls the newer positivist (or, perhaps better, naturalistic) approach. In this context, Hodge emerges as a commentator of great insight, one of the few who saw to the heart of Darwin's naturalistic project and realized that it represented a scientific *Weltanschauung* entirely incompatible with the Christian vision. In Gillespie's view it was Hodge, rather than Gray, who caught on first: "Despite his occasional prescience, it took Gray years to begin to appreciate the real threat contained in natural selection. Unlike Princeton theologian Charles Hodge who was, and remains, one of the most astute writers on the theological implications of Darwin's work, Gray long hoped that Darwin could and would declare natural selection and design compatible."[50] Another historian, Walter

---

Hodge's larger intentions, perpetuate confusion by not differentiating the issues of biblical interpretation (a secondary question for Hodge) and teleology in nature (the primary matter in Hodge's book). See, for example, John Dillenberger, *Protestant Thought and Natural Science: A Historical Interpretation* (Garden City, N.Y.: Doubleday, 1960), 236–45; Cynthia E. Russett, *Darwin in America: The Intellectual Response, 1865–1912* (San Francisco: W. H. Freeman, 1976), 26; and Gregory, "Impact of Darwinian Evolution," 377–78.

49. Paul F. Boller, Jr., *American Thought in Transition: The Impact of Evolutionary Naturalism, 1865–1900* (Chicago: Rand McNally, 1969), 22–23. Similar views are expressed by A. Hunter Dupree, *Asa Gray, 1810–1888* (Cambridge, Mass.: Harvard University Press, 1959), 360–63, 375; and Frank H. Foster, *The Modern Movement in American Theology* (New York: Revell, 1939), 48–50.

50. Neal C. Gillespie, *Charles Darwin and the Problem of Creation* (Chicago: University of Chicago Press, 1979), 112.

Wilkins, makes a similar claim in even stronger terms: "In contrast to some of his contemporaries, Hodge does understand Darwin's use of natural selection: natural laws working without design or final cause. As we have seen, natural selection without divine teleology is the key to Darwin's theory, and Hodge focused on this issue in the origin of species. . . . In spite of the rhetoric, which is all some critics have heard, Hodge did focus on the primary epistemological and methodological issues of Darwinism."[51]

Modern judgments like Gillespie's and Wilkins's reaffirm, though from a new point of view, what at least some of Hodge's contemporaries also held. Francis Bowen, a Unitarian philosopher at Harvard, was probably the country's most widely celebrated author of collegiate texts on logic, ethics, and political philosophy in the middle third of the century. His response (1860) to *The Origin of Species* was even more caustic than Hodge's. In defending the concept of final causes Bowen was convinced that even natural selection was in principle too much, since "the results of the Theory [of natural selection] are necessary or fatalistic; they blot God out of creation everywhere."[52] Hodge's advance over Bowen was to refine the definition of Darwinism to its ateleology and to leave the question of natural selection open to the results of empirical research.

An incident occurring only months after the publication of *What Is Darwinism?* showed why Hodge's conclusion that "Darwinism" simply means ateleology has carried considerable weight with at least some modern historians. When the British Association for the Advancement of Science, the leading scientific body of its day, held its annual meeting at Belfast in August 1874, the presidential address was given by John Tyndall, an aggressive self-styled Darwinian who used the occasion to lambaste all who were trying to incorporate principles of design into Darwin's theories. To refute what they took as an open promotion of atheism, several Belfast ministers made written protests. Robert Watts, a professor of theology at the Presbyterian Assembly's college who had studied at Princeton Seminary in the 1850s, took up some of the same arguments that Hodge had used only shortly before. When Watts sent to Princeton a copy

---

51. Walter J. Wilkins, *Science and Religious Thought: A Darwinian Case Study* (Ann Arbor: UMI Research Press, 1987), 44.
52. Francis Bowen, "The Latest Form of the Development Theory," in idem, *Gleanings from a Literary Life, 1838–1880* (New York: Scribner, 1880), 215.

of his response, Hodge was pleased with the substance of Watts's work. He was also heartened that the Darwinians, as he put it, "have at last got courage to speak out" in confessing frankly the atheistic bent of Darwinism.[53]

The judgment of Bowen and Watts, Hodge's contemporaries, supports the judgment of modern historians like Gillespie and Wilkins that Hodge had in fact caught the gist of Darwinism. This unusual array of commentators, while praising Hodge, share a viewpoint with those who dismiss *What Is Darwinism?* for failing to grasp Darwin's achievement. That common viewpoint is the conviction that "Darwinism" has a hard-edged clarity; that is to say, there is a simple essence to "Darwinism." One's particular understanding of that essence determines whether Hodge is viewed as advanced or retrograde.

4. A fourth category of critics disputes the assumption that a clear-edged definition of "Darwinism" actually exists. They respect Hodge for his efforts, but focus attention not so much on the essence of "Darwinism" as on how Hodge tried to penetrate through side issues to deal with the truly fundamental question. Hodge's contribution to the debates over religion and evolution is that he tried to unpack the question itself rather than racing to embroider an answer determined before the question was even asked. Among the critics in this camp are David Lindberg and Ronald Numbers:

> The conflicts surrounding Darwin were far more complex than the science-versus-religion formula suggests. They arose between persons who wished to retain an older, theologically grounded view of science, and those who advocated a thoroughly positivistic science; scientists as well as clerics could be found on each side, neither of which was motivated solely by scientific considerations. In contrast with the stereotypical view of Darwin's clerical critics as ignorant obscurantists, the Reverend Charles Hodge, who equated Darwinism with atheism, appears . . . as a man who saw the issues as clearly as Darwin. For such orthodox Christians the fundamental question was not the

---

53. Hodge to Watts, 12 October 1874, as printed in Robert Watts, *Atomism: Dr. Tyndall's Atomic Theory of the Universe Examined and Refuted. To Which Are Added . . . a Note from the Rev. Dr. Hodge . . .* (Belfast: William Mullan, 1875), 34. For a full examination of the confessional Presbyterian response to Tyndall's address, see David N. Livingstone, "Darwinism and Calvinism: The Belfast-Princeton Connection," *Isis* 83 (1992): 408–28.

incompatibility of Darwinism with the Mosaic account of creation nor the prospect of kinship with apes but the negative consequences of Darwinism for theism.[54]

Hodge was exceptional in that he realized that the issues at stake were neither simple nor self-evident.

Does Darwinism Have an Essence?

Our last category of scholars questions the basic assumption that there is an essence to Darwinism that can be used to categorize Darwinian friends and foes. Although both Hodge's book and many historical comments thereon share this assumption, it has been cast into doubt by some historians of evolutionary theory. James Moore, for example, has recently shown that already in the decade of the 1860s at least five different constructions of "Darwinism" were competing against each other. Moore comments: "The hermeneutic options are messy enough to make a tidy-minded scholar want to take sides. After all, who got it right? Who correctly interpreted what Darwin said? Who understood what Darwin really meant? Who has fair claim to represent authentic Darwinism? I find these questions unhistorical, and thus uninteresting."[55]

Moore's skepticism about "essential" Darwinism arises from the contestable nature of the "definitive" candidates that were advanced in the 1860s and that have been proposed since that decade. For Hodge, Darwin's antiteleological naturalism constituted the essence of the theory. Today, ironically, something akin to Hodge's definition can be found at both poles of the ideological spectrum.

---

54. Lindberg and Numbers, *God and Nature*, 14. Similarly nuanced views of Hodge's effort are found in Ian G. Barbour, *Issues in Science and Religion* (Englewood Cliffs, N.J.: Prentice-Hall, 1966), 98; Paul K. Conkin, *American Christianity in Crisis* (Waco, Tex.: Baylor University Press, 1981), 32–33 (though Conkin errs when he suggests that Hodge considered Darwin a materialist [p. 32]; see *What Is Darwinism?* 152); John C. Greene, "Darwinism as a World View," in idem, *Science, Ideology, and World View: Essays in the History of Evolutionary Ideas* (Berkeley: University of California Press, 1981), 151; John H. Brooke, *Science and Religion: Some Historical Perspectives* (Cambridge: Cambridge University Press, 1991), 303–5; and especially James R. Moore, *The Post-Darwinian Controversies: A Study of the Protestant Struggle to Come to Terms with Darwin in Great Britain and America, 1870–1900* (Cambridge: Cambridge University Press, 1979), 203–5, 211–12.

55. James R. Moore, "Deconstructing Darwinism: The Politics of Evolution in the 1860s," *Journal of the History of Biology* 24 (1991): 358.

For example, Richard C. Lewontin, who is committed to a naturalistic view of the world, sees the essence of Darwinism as "the replacement of a metaphysical view of variations among organisms by a materialistic view."[56] From the other side, Phillip Johnson, a critic of evolutionary cosmology, defines Darwinism as "fully naturalistic evolution—meaning evolution that is not directed or controlled by any purposeful intelligence."[57]

The difficulty with such judgments is that they do not contribute to historical understanding. On this reading, true Darwinians were as rare as gold dust in the nineteenth century. With this definition—provided, as it were, by the contemporary polarities—such key figures in the nineteenth century as Asa Gray, Charles Lyell, George Henslow, and Alfred Russel Wallace could not be considered Darwinians, even though all of them thought of themselves as Darwinians.[58] Moreover, according to Robert Richards, virtually the entire tradition of evolutionary psychology in the nineteenth century was constructed by researchers who took their cues from Darwin, but who also maintained metaphysical (or even more directly religious) commitments.[59] To rule on simple definitional grounds that they were not "Darwinians" would surely be misguided.

The problem is not merely that the definition in view is inadequate. An alternative suggestion—that the essence of Darwinism lies in Darwin's idea of *gradual* species transformation—rules out such a key Darwinian as Thomas Henry Huxley, who favored saltationism (the notion that species are transformed through dramatic leaps). Moreover, contemporary advocates of punctuated equilibrium—a conception of evolutionary change that sees organic transformation

---

56. Cited in David L. Hull, *Science as a Process: An Evolutionary Account of the Social and Conceptual Development of Science* (Chicago: University of Chicago Press, 1988), 774.

57. Phillip E. Johnson, *Darwin on Trial* (Washington, D.C.: Regnery Gateway; and Downers Grove, Ill.: Inter-Varsity, 1991), 4n.

58. For biographical sketches of Lyell, Henslow, and Wallace see p. 76, n. 21; p. 91, n. 40; and p. 133, n. 94.

59. Robert J. Richards, *The Meaning of Evolution: The Morphological Construction and Ideological Reconstruction of Darwin's Theory* (Chicago: University of Chicago Press, 1992). In light of persistent efforts in the nineteenth century to give Darwin's theories a teleological cast, Peter J. Bowler has argued that there simply was no Darwinian revolution until the neo-Darwinian synthesis of the 1930s when random mutations and biological materialism began to achieve the status of scientific orthodoxy (*The Non-Darwinian Revolution* [Baltimore: Johns Hopkins University Press, 1988]).

proceeding by fits and starts—such as Stephen Jay Gould and Niles Eldredge would not pass muster as good Darwinians.

Most important, restricting Darwinism to mean evolution exclusively by means of natural selection faces the embarrassing obstacle that it would rule out Darwin as a Darwinian, since he allowed for evolution by family selection, use inheritance, sexual selection, and correlative variation.[60] Besides, Darwin also obviously changed his mind on whether his system could be made compatible with design. Shortly after publishing the *Origin,* Darwin himself footed the bill for distributing in Britain several of Asa Gray's articles that had been gathered into a pamphlet called *Natural Selection Not Inconsistent with Natural Theology.* Yet by 1864 Darwin apparently had second thoughts and so stopped promoting this pamphlet.[61]

Given the difficulties in assigning scientists, theologians, and cosmologists to neat Darwinian or anti-Darwinian categories, it is not surprising to find a modern historian, David Hull, arguing that there is no essence to Darwinism at all, and that in writing the history of Darwinian theory we would be better off simply looking at those individuals who considered themselves Darwinian.[62] This proposal, of course, moves the focus of debate away from narrowly cognitive or conceptual claims towards broader social and intellectual factors. Because it suggests that the very notion of Darwinism is a historical construct constituted and transformed by the members of Darwinian and anti-Darwinian circles themselves, any study of Darwinism must, by its nature, also (1) investigate how Darwinian ideas relate to other ideas in a particular culture, (2) examine who benefits and who loses when a particular definition of Darwinism triumphs, and (3) realize that the intellectual predispositions of com-

---

60. Ernst Mayr, "Darwin's Five Theories on Evolution," and William B. Provine, "Application and Mechanisms of Evolution after Darwin: A Study in Persistent Controversies," in *The Darwinian Heritage,* ed. David Kohn (Princeton: Princeton University Press, 1985), 755–72, 825–66.

61. See Desmond and Moore, *Darwin,* 502, 526; David N. Livingstone, "The Idea of Design: The Vicissitudes of a Key Concept in the Princeton Response to Darwin," *Scottish Journal of Theology* 37 (1984): 333; and for a general discussion of Darwin's shifting views see Dov Ospovat, "God and Natural Selection: The Darwinian Idea of Design," *Journal of the History of Biology* 13 (1980): 169–94. There is a good discussion of how Asa Gray's views on teleology differed from Darwin's in Phipps, "Asa Gray's Theology of Nature," 170.

62. Hull, *Science as a Process.*

mentators are everywhere shaping their conclusions about the objective data of science.

We can apply this broader cultural viewpoint to our evaluation of Hodge. Clearly Hodge would have been very unhappy—unlike, say, Gray, McCosh, or even Warfield—to be considered a Darwinian. (At least at one point in his career, Warfield described himself as "a Darwinian of the purest water.")[63] Darwinians, as Hodge defined them, moved in a world of naturalistic explanation that Hodge found repelling; they shared a philosophical perspective that was inimical to his own. Perhaps this was the very reason why his book was so conspicuously unconcerned with biblical exegesis, but so profoundly exercised over philosophical-theological concerns. By contrast, later bearers of the Princeton tradition had grown accustomed to the scientific vision, so much so that Warfield believed Christians could hold a virtually mechanistic account of nature provided they allowed for at least occasional supernatural intervention and maintained the belief that providence sustains the physical world. Thus it was not so much that Hodge had a poorer or clearer understanding of Darwinian theory than did, say, his son A. A. Hodge or Warfield. Rather, by the time of these later theologians the socioscientific ethos had changed. The scientific enterprise and its professionalized champions had come to stay, and the need to work out a theological modus vivendi seemed more necessary than ever.

These observations are not meant to question the wisdom of Hodge's interpretative efforts. They are merely to show that a broader framework is helpful when considering how Darwinism was (or should be) constituted as a historical entity. In sum, Hodge may have been correct that what he defined as Darwinism was incompatible with orthodox Christianity. But others may also have been within their epistemic rights to consider themselves Darwinians even while holding to a teleological vision.

Teleology and Natural Theology

If Hodge's conception of Darwinism is worthy of our contemplation, no less so is his judgment concerning the role of teleology in Christian dogma. Hodge thought it was self-evident that a denial of design meant a denial of Christianity. To Hodge this truth was so

---

63. B. B. Warfield, "Personal Reflections of Princeton Undergraduate Life," *Princeton Alumni Weekly*, 19 April 1916, p. 652.

Introduction: Charles Hodge and the Definition of "Darwinism"

obvious that he did not even see the need to discuss the subject in *What Is Darwinism?* where teleology is so important. But just what role did teleology play in Hodge's theology?

In the late twentieth century it has become a commonplace to implicate advocates of teleology in some form of natural theology, which is the attempt to deduce the existence and (to some extent) the character of the Creator from scrutinizing the works of creation. Such efforts have been the subject of manifold attacks in the twentieth century, sometimes from theological liberals, but even more often from neoorthodox opponents of liberalism or philosophical defenders of traditional Protestantism.[64] The basic criticisms are that natural theology gives too much credit to the human ability to fathom God's purposes in nature, relies too much on the thought forms of the particular age in which its arguments are constructed, and identifies too casually the theologian's perspective with the divine perspective. If these criticisms apply to Hodge's teleology as well, the baseline of his attack on Darwinism would be seriously compromised.[65]

The notion that Hodge's teleology falls prey to the problems of natural theology has been recently questioned by Jonathan Wells in his monograph *Charles Hodge's Critique of Darwinism*. Wells argues that Hodge's writings show little concern for traditional natural theology in the Thomistic sense; indeed, he cites Hodge to the effect that standard proofs for the existence of God are "not designed so much to prove the existence of an unknown being, as to demonstrate that the Being who reveals himself to man in the very constitution of his nature must be all that Theism declares him to be."[66] Hodge's defense of design, then, arose from his prior commitment to the God revealed in Scripture. As Wells puts it, Hodge was less

---

64. Complaints about natural theology have been made by Emil Brunner, H. Richard Niebuhr, Abraham Kuyper, Cornelius Van Til, Nicholas Wolterstorff, Alvin Plantinga, and many other recent theologians and philosophers. A general discussion of some of these arguments is found in Bruce A. Demarest, *General Revelation: Historical Views and Contemporary Issues* (Grand Rapids: Zondervan, 1982).

65. For an idea of the reservations that such scruples concerning natural theology have raised about the work of the Princeton theologians, see John C. Vander Stelt, *Philosophy and Scripture: A Study in Old Princeton and Westminster Theology* (Marlton, N.J.: Mack, 1978).

66. Jonathan Wells, *Charles Hodge's Critique of Darwinism: An Historical-Critical Analysis of Concepts Basic to the 19th Century Debate* (Lewiston, N.Y.: Edwin Mellen, 1988), 40.

concerned about the standard argument *from* design than with asserting an argument *to* design. In the former case, truths about God are supposedly deduced inferentially from observed design in the world; in the latter, belief in a designed world is the outcome of a prior belief in Scripture. Accordingly, when Hodge refers his readers to such classics of natural theology as William Paley and the Bridgewater Treatises, his purpose is less to adduce empirical evidence for the existence of God than to present demonstrations of the qualities already known about God from Scripture.

If Wells's interpretation is correct, Hodge ought not be seen as a defender of traditional natural theology; rather, he was a defender of design simply because design is part of the very confessional fabric of Christianity. With such a conviction, there was all the more reason for Hodge to assert that Darwinism was atheism, pure and simple. Yet Wells's evaluation, though it advances a most interesting argument, sits uncomfortably with Hodge's own observation that the Darwinian expulsion of design from nature brought "it into conflict not only with Christianity, but with the fundamental principles of natural religion" (p. 92). At least some of the arguments in *What Is Darwinism?* seem to lean more heavily on what Hodge took to be the evidences of nature ("the fundamental principles of natural religion") than Wells's interpretation would suggest.

Whether Hodge was quite so dedicated to constructing an argument *to* design as Wells would have us believe, it is certainly clear that the idea of design was of crucial importance in his judgments. Moreover, the history of Hodge's wrestling with Darwinism reveals how central the teleological vision is to doctrinally diverse Christians. Calvinists with anxieties about natural theology, and Wesleyans who welcomed it, were both likely to hesitate over any system of organic history that relegates design to irrelevance.[67] For both, God's design was fundamental to the doctrinal fabric of Christianity.

Hodge's Notion of Design

The last matter that calls for precise definition is Hodge's particular conception of design, which, as we have seen, was his key in

---

67. For more extensive discussions see David N. Livingstone, *Darwin's Forgotten Defenders: The Encounter between Evangelical Theology and Evolutionary Thought* (Grand Rapids: Eerdmans; and Edinburgh: Scottish Academic Press, 1987), 100–45; and Livingstone, "Idea of Design."

interpreting Darwinism. Hodge's idea of design was in part the age-old Scripture-rooted Christian notion that "since the creation of the world God's invisible qualities—his eternal power and divine nature—have been clearly seen, being understood from what has been made" (Rom. 1:20). But it was also a product of distinctly particular contributions of eighteenth-century Christian apologists who hoped to maintain theological credibility in competition with the rise of early modern science.

As *What Is Darwinism?* shows, Hodge's conception of how God had designed the world had several distinct features that were, at least in part, specific to his age. His was a view of design in which there was little mystery: a right understanding of God's work in the world, said Hodge, quoting Walter Mitchell, "leaves no mysteries in the animate world unaccounted for" (p. 124). It was also a view of design in which dignity, defined in nineteenth-century Victorian terms, was crucial: "The dog which spends its life in snarling contention with its fellow curs for insufficient food, will not be a noble specimen of its race. God does not so treat his creatures" (p. 129). It was a view of design whose utilitarian conclusions had been earlier defined by William Paley: "To any ordinarily constituted mind, it is absolutely impossible to believe that [the eye] is not a work of design" (p. 96).[68] It was a view of design heavily influenced by nineteenth-century conceptions of the extraordinary power of commonsense intuitions: "It is ground of profound gratitude to God that He has given to the human mind intuitions which are infallible, laws of belief which men cannot disregard any more than the laws of nature" (p. 137). Or again, "In thus denying design in nature, these writers array against themselves the intuitive perceptions and irresistible convictions of all mankind—a barrier which no man has ever been able to surmount" (p. 153). Finally, it was a view of design heavily dependent upon an idealized Baconian science that

---

68. At this point in *What Is Darwinism?* Hodge quotes Darwin from the *Origin* as follows: "It is indispensable to arrive at a just conclusion as to the formation of the eye, that the reason should conquer the imagination; but I have felt the difficulty far too keenly to be surprised at any degree of hesitation in extending the principle of natural selection to so startling an extent." The burden of argument which Hodge began to take up was to show that natural human incredulity at the formation of the eye without teleological foresight was a fundamental, as opposed to merely conceptual, objection to ateleological natural selection. That Darwin himself felt the objection suggests something about the seriousness of Hodge's commonsense scruples.

had perhaps a greater influence on the theologians than on the working scientists of the era.[69]

Committed as he was to the particular shape that "design" had assumed over the preceding two centuries, Hodge may have confused Darwin's transgression of that particular notion of design with a denial of design per se. The sharpest student of such matters, James Moore, has praised Hodge for the depth of his perception in *What Is Darwinism?* but he has also gone on to ask if the notions of the certainty and the fixity of species that Hodge (and other shrewd anti-Darwinists like John William Dawson) brought against Darwin were not more the product of the early nineteenth century than of timeless Christian truth.[70]

In sum, it is imperative to take Hodge with great seriousness as he defines "Darwinism" and, on the basis of that definition, makes his final judgments. But it is just as important to notice how Hodge defined the "design" that provided him with an ultimate standard. Hodge's judgment that Darwinism amounted to atheism stemmed from very specific definitions of the essence of Darwin's theory and of design. Whether Hodge's diagnosis was sound, or indeed whether the whole project of identifying necessary and sufficient conditions for Darwinism can be sustained, there is little doubt that how Christians responded to Darwin's theory depended crucially on what they understood Darwinism to be. That principle holds today as well. Our response to Darwinism depends on our definitions of Darwinism and design. To simply assume that design is a self-evident concept that does not have to be explored with as much care as Hodge expended in exploring Darwin's ideas is to short-circuit the very project Hodge promoted. Whether we ally ourselves in the end with the camp that defines Darwinism as a purely naturalistic explanation of organic history, or with the camp that allows as Darwinians those who see some overarching purpose in evolu-

---

69. That at least is the conclusion of Richard Yeo, "An Idol of the Market-Place: Baconianism in Nineteenth Century Britain," *History of Science* 23 (1985): 251–98.

70. Moore, *Post-Darwinian Controversies*, 213–16. An American scientist and theologian who made at least some efforts to construct a fully Calvinistic idea of design based on Darwin's theories was George Frederick Wright, who also assisted Gray with several of his published writings on Darwin and teleology. See Livingstone, *Darwin's Forgotten Defenders*, 65–70; and Ronald L. Numbers, "George Frederick Wright: From Christian Darwinist to Fundamentalist," *Isis* 79 (1988): 624–45.

Introduction: Charles Hodge and the Definition of "Darwinism"

tionary theory, is a critical decision.[71] Either way, to control definition is to exert cultural power.

Five works, or excerpts of works, follow. We begin with a footnote to an article Hodge wrote in 1862; the footnote is his first published comment on Darwin's *Origin of Species*. Next are two statements showing Hodge's belief in the integrity of science and the need for students of Scripture to respect the findings of scientific research. Finally, *What Is Darwinism?* and Asa Gray's review of this work are presented in full. These last two documents are critical in that they illustrate the power of definition in shaping debate.

Together, these works contribute to a better understanding of what was at stake between alternative views of science in the years immediately after Darwin's famous publications. They also show how important it is for contemporary students to define with care not only the conclusions of scientists, but also the constructs by which scientists and theologians, like all human beings, make sense of what they discover in the world.[72]

---

71. Examples of those who have found design in evolutionary theory come from the four corners of the theological compass, including the Roman Catholic Pierre Teilhard de Chardin, the Orthodox Theodosius Dobzhansky, conservative Protestants like Donald MacKay, R. J. Berry, Douglas Spanner, and Howard Van Till, and many members of the American Scientific Affiliation.

72. Since the editors have stressed the way in which one's particular standpoint influences critical judgment, it is important that we declare ourselves as well. We hold to the traditional Christian view that the Bible is true, but we are skeptical about attempts to harmonize scientific conclusions with particular interpretations of passages in Scripture, including early Genesis. We are convinced that God is both the source of all the processes by which life came into existence and the sustainer of all those by which it continues to exist, but also that some of these processes may simply not be fathomable by humans. Epistemologically, we are moderate realists who hold that empirical investigations can in fact reveal some aspects of the truth about nature, but also that larger interpretations of the world are inevitably constructs defined by the social, political, religious, and cultural positions of those who maintain them.

Three Brief Notices

Hodge's First Published Comment on Darwin (1862)

Charles Hodge's first published notice of Darwin's *Origin of Species* appeared in the July 1862 issue of the *Biblical Repertory and Princeton Review* (vol. 34, no. 3). The occasion was an essay entitled "Examination of Some Reasonings against the Unity of Humankind" (pp. 435–64) in which Hodge objected to the increasing willingness in his day to postulate multiple origins for the human race. The burden of the article was to show that the preponderance of then-known scientific evidence spoke in favor of a unitary beginning for humanity, even though a few notable scientists argued to the contrary. Among such scientists whom Hodge took to task were Samuel George Morton, who in the 1830s and 1840s had used skull measurements to defend the view that humanity was made up of different *species;* Josiah Nott and George Gliddon, who likewise advocated polygenism (i.e., multiple origins for the human race); and Louis Agassiz, the famous Harvard naturalist who later would oppose Darwin even more vigorously than Hodge did. The essay testifies to Hodge's wide reading in the work of American and European naturalists, and also to his thorough grounding in different theological attempts to harmonize traditional biblical interpretations and the era's most conspicuous scientific proposals. Hodge's conclusion was that the picture of a unified humanity as found in the first chapter of Genesis—even considered "merely as a philosophical theory of the beginning of things, the result merely of a wise man's reflections, after a wide examination of the phenomena of nature and of man" (p. 463)—accorded much better with the facts of nature than did the notion of a humanity divided into several species.

Darwin enters the picture in a footnote. At the end of a long refutation of a work by Americans Nott and Gliddon (*Types of Mankind, or Ethnological Researches, Based upon the Ancient Monuments* [Philadelphia, 1854]), Hodge cites a book by the British naturalist Richard Owen (*On the Classification of the Mammalia* [London, 1859]) which supports his conclusion about the unity of humanity. Owen was a leading British naturalist of his day who strategically combined natural theology and French anatomical morphology to promote a teleological view of nature different from William Paley's. Owen's scheme featured the idea of "homology": a range of archetypal plans originally created by God persist throughout the created order. Owen's skepticism about Darwin's arguments in *The Origin of Species* was for several years the most formidable British obstacle to their acceptance. Having summarized several arguments by Owen against the specific contentions of Nott and Gliddon, Hodge quotes from Owen's book: "Thus, I trust, has been furnished the confutation of the notion of the transformation of the ape into man." At that point (p. 461) appears the following footnote concerning Darwin's *Origin,* which Hodge, despite his preoccupation with the American Civil War and a multitude of more narrowly ecclesiastical and educational duties, seems already to have read, weighed, and found wanting.

As we have seen, it was no part of the design of the authors of the *Types [of Mankind]* to advocate the same origin, or unity of species, for man and the monkey. This belongs to the opposite pole of sceptical speculation in natural history, of which the latest form appears in a remarkable book from a very high authority: *On the Origin of Species,* &c.; by Charles Darwin, M.A., Fellow of the Royal, Geological, Linnaean, &c., Societies, 1860. The object of this interesting work is to prove that there is no such thing as permanence in the species of natural history, that all existing forms of animal life have been derived through natural generation from one or at most a very few original creations. It carries, however, its own refutation in itself in the author's frank admission of the difficulties of his theory, and in the stupendous absurdity of his conclusion. This is expressed as follows: "I believe that animals (i.e., all animals) have descended from at most only four or five progenitors, and plants (all) from an equal or lesser number. . . . I should infer, from analogy, that probably all the organic beings which have ever lived on this earth, have descended from some one primordial form, into which life was first breathed." Cuvier[1] has characterized, for all time, this whole branch of speculation in the brief words: "There is no proof that all the differences which now distinguish organized beings are such as may have been produced by circumstances; all that has been advanced upon this subject is hypothetical." Since his day, however, these speculations, even of the greatest authorities within the legitimate sphere of the science, have become mutually self-destructive to a degree which Cuvier could never have anticipated. Thus Morton and Agassiz find such differences between man and man that the different races or groups never could have descended from a single pair, while Darwin finds so little difference between man and the animals that he believes them all to be "descended from at most only four or five progenitors," and infers "from analogy" that they are all "descended from some one primordial form." It is quite certain that such conflicting conclusions cannot endanger the received doctrines of the immutable permanency of species, and of the specific unity of the human race.

Response to the *New York Observer* (1863)

The January 1863 issue of the *Biblical Repertory and Princeton Review* published an essay entitled "The Skepticism of Science." That article, by a widely read minister-businessman from Pennsylvania, Joseph Clark (1825–65), generated a sharp rebuttal from the editor of the *New York Observer,* a leading periodical for conservative Congregationalists. Charles Hodge, in turn, responded to that criticism by presenting a brief, clear statement of his own opinions about the relationship between the Bible and science.

In his essay, Clark began by suggesting that because of rapid development in the sciences "the great battles of Christianity henceforth are to be fought with the various forms of unbelief gen-

---

1. Baron Georges Cuvier (1769–1832) originated a system of animal classification and contributed to the founding of vertebrate paleontology. To explain the succession of fossils he discovered, he resorted to the theory of catastrophism rather than evolutionary process.

erated by scientific inquiry."[2] His strategy for dealing with this problem was twofold: first, maintain the full trustworthiness of the Bible; but, second, allow scientists to pursue the proper inductive procedures of their disciplines without precommitments to conclusions thought to be found in the Bible. Clark pointed out that earlier interpretations of Scripture (like the belief that the Bible teaches a flat earth) had been revised in light of modern science. Although further reinterpretations of that kind might be necessary, no damage would result to the church and to the doctrine of biblical inspiration if believers allowed scientists to pursue their investigations, and if Christians themselves became better acquainted with the procedures and results of modern science. Clark concluded that the tensions between science and religion were more apparent than real.

These concessions to the sphere of science proved too much for the *New York Observer.* In a sharp rejoinder in the March 12, 1863, issue, the editor stoutly defended traditional interpretations (e.g., humanity is six thousand years old), and also pictured the clash between religion and science as much more fundamental than had the Reverend Clark: "If we are to accept the teachings of *science* and set *revelation* aside when the deductions of the former conflict with the latter, where are we?" The editor felt that the *Princeton Review* article amounted to "a distinct declaration that in *scientific* matters, faith in Moses and Paul must yield to [the modern scientists] Humboldt and Agassiz." There was, the editor countered, no reason to change traditional interpretations in response to modern science: "Heaven and earth may pass away, but the Word of the Lord standeth sure. And science has never found anything yet in the heavens or the earth to shake a good man's faith in the accuracy of the sacred record of their creation."[3]

Hodge's response (under the heading "The Bible in Science") came two weeks later in the March 26 issue of the *New York Observer* (pp. 98–99). It reflected his long-standing efforts to maintain harmony between divine truth expounded in Scripture and valid

---

2. [Joseph Clark], "The Skepticism of Science," *Biblical Repertory and Princeton Review* 35 (Jan. 1863): 43. For identification of Clark as the author of this article, which, like all others in the journal, was published anonymously, see *Biblical Repertory and Princeton Review: Index Volume from 1825 to 1868*, 126–27.

3. "Scripture and Science," *New York Observer,* 12 March 1863, p. 82. On the *New York Observer,* see Charles H. Lippy, ed., *Religious Periodicals of the United States* (Westport, Conn.: Greenwood, 1986), 169.

knowledge of the natural world learned from scientific procedures. Thus, on the one hand, Hodge offered an even stronger statement defending the inerrancy of the Bible than that found in his *Systematic Theology,* where he conceded that there might be "a speck of sandstone" in the marble of the biblical Parthenon.[4] But, on the other hand, Hodge also spoke out vigorously for the rights of inductive science and for the possibility that scientific discoveries might require new interpretations of Scripture.

To the editors of the *New York Observer:*

Your paper has recently published two notices of the last number of the *Princeton Review,* neither of which, as it seems to me, evinces your usual discrimination or justice. [The first notice attacked the *Princeton Review*'s defense of Abraham Lincoln's suspension of habeas corpus during the Civil War.] . . .

Believing, however, that the article in question [on President Lincoln] might safely be allowed to speak for itself, the Editor of the *Review* would not have deemed it necessary to trouble your readers with this communication had you not brought against another article in the same number the far more serious charge of abetting infidelity. You say in your paper of March 12th that the reviewer whose curious concessions to skepticism called forth your remarks lays down certain propositions which you assert to be "a distinct declaration that in *scientific* matters, faith in Moses and Paul must yield to Humboldt and Agassiz." It would not be for edification to answer your remarks in detail. It will be better to state in few words what has always been the ground occupied by the *Princeton Review* in reference to the relation between Science and Revelation. In the first place, the *Review* and its conductors have always asserted the plenary inspiration of the Bible, in such sense that what the sacred writers say, the Holy Ghost said. Their words are his words. From this it follows that the Bible can teach no error, whether in reference to doctrines, morals, or facts; whether those facts be historical, geographical, geological or astronomical.

In the second place, however, the *Princeton Review* has ever held and taught, in common with the whole Church, that this infallible Bible must be interpreted by science. But what is science? It is a pity

---

4. Charles Hodge, *Systematic Theology,* 3 vols. (New York: Scribner, 1872–73), 1:169–72.

you did not ask yourself that question before saying that, according to the reviewer, "faith in Moses and Paul must yield to Humboldt and Agassiz." You surely do not need to be told that the dicta, the theories, the opinions, the vagaries of no one man, or number of men, are science. These are unreliable, inconstant, and inconsistent. You could hardly have hit on two distinguished men whose scientific opinions on several vital points are more antipodal than Humboldt and Agassiz.[5] Science is not the opinions of man, but knowledge; and specially, according to usage, the ascertained truths concerning the facts and laws of nature. To say, therefore, that the Bible contradicts science is to say that it contradicts facts, is to say that it teaches error; and to say that it teaches error is to say it is not the Word of God. The proposition that the Bible must be interpreted by science is all but self-evident. Nature is as truly a revelation of God as the Bible, and we only interpret the Word of God by the Word of God when we interpret the Bible by science. As this principle is undeniably true, it is admitted and acted on by those who, through inattention to the meaning of terms, in words deny it. When the Bible speaks of the foundations or of the pillars of the earth, or of the solid heavens, or of the motion of the sun, do not you and every other sane man interpret this language by the facts of science? For five thousand years the Church understood the Bible to teach that the earth stood still in space, and that the sun and stars revolved around it. Science has demonstrated that this is not true. Shall we go on to interpret the Bible so as to make it teach the falsehood that the sun moves round the earth, or shall we interpret it by science and make the two harmonize? Of course, this rule works both ways. If the Bible cannot contradict science, neither can science contradict the Bible. All those theories which in former ages scientific men have advanced against the clear teaching of the Scriptures have come to shame, and now every enlightened Christian votary of science knows that if his investigations seem to lead

---

5. Louis Agassiz organized his biogeographical interpretations around a belief in several distinct centers of creation. In sharp contrast to Agassiz, Alexander von Humboldt believed that there was only one form of human being, but made no effort to explain the harmony of the cosmos by reference to a divine Creator. For a brief biographical sketch of Agassiz, see p. 113, n. 70. Humboldt (1769–1859) was a Prussian naturalist who stressed the importance of investigating nature empirically and who did not feel constrained to interpret geological results within the context of a literal interpretation of Genesis.

to conclusions contrary to the Bible, there must be some error in his process. The diversity among the races of men, for example, is an undeniable fact. This diversity may be accounted for by ascertaining different origins for the several races, or by the influence of climate and modes of life.[6] As the Bible asserts the historical as well as the specific unity of the race, there is an end to the question. There is a two-fold evil on this subject against which it would be well for Christians to guard. There are some good men who are much too ready to adopt the opinions and theories of scientific men, and to adopt forced and unnatural interpretations of the Bible, to bring it to accord with those opinions. There are others who not only refuse to admit the opinions of men, but science itself, to have any voice in the interpretation of Scripture. Both of these errors should be avoided. Let Christians calmly wait until facts are indubitably established, so established that they command universal consent among competent men, and then they will find that the Bible accords with those facts. In the meantime, men must be allowed to ascertain and authenticate scientific facts in their own way, just as Galileo determined the true theory of the heavens.[7] All opposition to this course must be not only ineffectual, but injurious to religion. They do not fear that science will ever subvert the moral law, because they have faith in that law. If they had like faith in the

---

6. At this point in his argument, Hodge may well have been recalling the arguments of Samuel Stanhope Smith of the College of New Jersey, who published an expanded version of his *Essay on the Causes of the Variety of Complexion and Figure in the Human Species* in 1810, shortly before Hodge began his study at the college. Smith had argued, precisely as Hodge summarizes here, that climate and cultural differences are adequate to explain differences among human types. On this work and its reflection of the Baconian theism that was always so important to Hodge, see Mark A. Noll, *Princeton and the Republic, 1768–1822* (Princeton: Princeton University Press, 1989), 115–24, 186–88, 196–98.

7. With his Baconian notions of science, Hodge misstates slightly what had actually occurred. Galileo's empirical observations had not so much proved the heliocentric system as they had disproved Ptolemy's geocentricism. In Galileo's own era, the Danish astronomer Tycho Brahe (1546–1601) had shown coherently how Galileo's observations could be used to support a cosmology later known as Tychonism or geoheliocentricism, according to which the planets revolve around the sun, which itself revolves around the stationary earth. For the nature of Tycho's reasoning, see Robert S. Westman, "The Copernicans and the Churches," in *God and Nature: Historical Essays on the Encounter between Christianity and Science,* ed. David C. Lindberg and Ronald L. Numbers (Berkeley: University of California Press, 1986), 84–85.

Bible, they would feel assured that it will stand, as it has ever stood, impregnable as truth.

Very truly your friend,
The Editor of the *Princeton Review*

Excerpts from the *Systematic Theology* (1872–73)

The best-known statement of the views that were taught at Princeton Seminary for over a century by Charles Hodge along with his predecessors and successors was Hodge's own *Systematic Theology*. This three-volume work of well over two thousand pages was published in 1872 and 1873, near the end of Hodge's active career as a teacher. Its concern for the reconciliation of biblical theology and the results of science maintained a distinctive tradition that had marked seminary education at Princeton since its beginning under Archibald Alexander.

At the start of the *Systematic Theology* Hodge constructed an analogy between the work of the scientist and the work of the theologian. In so doing he drew upon the confidence in Baconian induction that dominated American thinking for most of the nineteenth century. "The Bible," wrote Hodge, "is to the theologian what nature is to the man of science. It is his store-house of facts, and his method of ascertaining what the Bible teaches is the same as that which the natural philosopher adopts to ascertain what nature teaches" (1:10). Thereafter, although the *Systematic Theology* was mostly occupied with treating standard biblical themes, it also took many pages to examine questions at the intersection of science and theology. Thus an extensive chapter on "Creation" dealt with what were then the most recent opinions on the origins of the earth (1:550–74), and another chapter on the "Unity of the Human Race" rehearsed Hodge's defense from the *Princeton Review* of a unified human species (2:77–91). In addition, most of Hodge's chapter on the "Origin of Man" was devoted to a broad consideration of Darwin's writings (2:3–41). As he would later repeat in *What Is Darwinism?* Hodge contended that Darwin's theory, especially its argument "against any divine intervention in the course of nature, and especially in the production of species" (2:17), was atheistic. At the same time, Hodge's treatment of Darwin in the *Systematic Theology* was more general and more rhetorical than the detailed discussion in *What Is Darwinism?*

The other side of the Bible-and-science picture in the *Systematic Theology* was Hodge's deference to science. Repeatedly he stressed that the work of scientists, as also philosophers, should be used to help theologians interpret the Scriptures. In making this assertion, Hodge developed more systematically the argument he had earlier presented to the editor of the *New York Observer*.

What follows is three separate paragraphs from the *Systematic Theology* in which Hodge spells out the way in which scientific findings should help interpret the Bible. The first is from an early section entitled "Relation of Philosophy and Revelation" (1:59), the second from a consideration of "Historical and Scientific Objections" to the doctrine of the plenary inspiration of Scripture (1:170–71), and the third is the concluding paragraph of his discussion "Geology and the Bible" (1:573–74). On that subject Hodge had proposed two ways of "reconciling the Mosaic account [in Genesis] with those facts" that geologists then seemed agreed upon concerning the great age of the earth (1:570). These were, first, the postulation of an immense time span between the original creation (Gen. 1:1) and the specific formation of the world (Gen. 1:2 and following), and, second (the position that Hodge favored), the hypothesis that each of the six "days" of creation stood for lengthy periods of geological time. As it happens, these views were also held by some of the most respected orthodox authorities on the Bible and science during the first half of the nineteenth century. Scotland's influential Thomas Chalmers (1780–1847), for example, favored the gap theory, while his contemporary and countryman Hugh Miller (see p. 148, n. 114) held the day-age theory. Significantly, it was Arnold Guyot (1807–84) of the College of New Jersey who popularized the day-age theory in America. James Dana of Yale (see p. 144, n. 105) also became an influential advocate of this theory.

Details aside, the most critical point Hodge makes in his treatment of geology and other knotty questions involving the Bible and science is his insistence that theologians should treat scientific results with all the respect possible within the general framework of faithfulness to Scripture.

The relation ... between philosophy and revelation, as determined by the Scriptures themselves, is what every right-minded man must approve. Everything is conceded to philosophy and science, which they can rightfully demand. It is admitted that they have a large and important sphere of investigation. It is admitted that within that sphere they are entitled to the greatest deference. It is cheerfully conceded that they have accomplished much, not only as means of mental discipline, but in the enlargement of the sphere of human knowledge, and in promoting the refinement and well-being of men. It is admitted that theologians are not infallible in the interpretation of Scripture. It may, therefore, happen in the future, as it has in the past, that interpretations of the Bible, long confidently received, must be modified or abandoned to bring revelation into harmony with what God teaches in his works. This change of view as to the true meaning of the Bible may be a painful trial to the Church, but it does not in the least impair the authority of the Scriptures. They remain infallible; we are merely convicted of having mistaken their meaning. . . .

The second great objection to the plenary inspiration of the Scripture is that it teaches what is inconsistent with historical and scientific truth. [The first objection was that the Scripture writers contradict each other.]

Here again it is to be remarked (1) That we must distinguish between what the sacred writers themselves thought or believed, and what they teach. They may have believed that the sun moves round the earth, but they do not so teach. (2) The language of the Bible is the language of common life, and the language of common life is founded on apparent, and not upon scientific truth. It would be ridiculous to refuse to speak of the sun rising and setting because we know that it is not a satellite of our planet. (3) There is a great distinction between theories and facts. Theories are of men. Facts are of God. The Bible often contradicts the former, never the latter. (4) There is also a distinction to be made between the Bible and our interpretation. The latter may come into competition with settled facts, and then it must yield. Science has in many things taught the Church how to understand the Scriptures. The Bible was for ages understood and explained according to the Ptolemaic system of the universe; it is now explained, without doing the least violence to its language, according to the Copernican system. Christians have commonly believed that the earth has existed only a few thousands of years. If geologists finally prove that it has existed for myriads of ages, it will be found that the first chapter of Genesis is in full accord with the facts, and that the last results of science are embodied on the first page of the Bible. It may cost the Church a severe struggle to give up one interpretation and adopt another, as it did in the seventeenth century, but no real evil need be apprehended. The Bible has stood and still stands in the presence of the whole scientific world with its claims unshaken. Men hostile or indifferent to its truth may, on insufficient grounds, or because of their personal opinions, reject its authority; but even in the judgment of the greatest authorities in science, its teachings cannot fairly be impeached. . . .

As the Bible is of God, it is certain that there can be no conflict between the teachings of the Scriptures and the facts of science. It is not with facts, but with theories, believers have to contend. Many such theories have, from time to time, been presented, apparently or really inconsistent with the Bible. But these theories have either proved to be false or to harmonize with the Word of God, properly interpreted. The Church has been forced more than once to alter her interpretation of the Bible to accommodate the discoveries of science. But this has been done without doing any violence to the Scriptures or in any degree impairing their authority. Such change, however, cannot be effected without a struggle. It is impossible that our mode of understanding the Bible should not be determined by our views of the subjects of which it treats. So long as men believed that the earth was the centre of our system, the sun its satellite, and the stars its ornamentation, they of necessity understood the Bible in accordance with that hypothesis. But when it was discovered that the earth was only one of the smaller satellites of the sun, and that the stars were worlds, then faith, although at first staggered, soon grew strong enough to take it all in and rejoice to find that the Bible, and the Bible alone of all ancient books, was in full accord with these stupendous revelations of science. And so if it should be proved that the creation was a process continued through countless ages, and that the Bible alone of all the books of antiquity recognized that fact, then, as Professor Dana says, the idea of its being of human origin would become "utterly incomprehensible."

What Is Darwinism?

The text that follows reproduces Hodge's *What Is Darwinism?* as published by Scribner, Armstrong, and Company of New York in 1874 and by T. Nelson and Sons of London and Edinburgh the same year. The only difference between the two is that the British edition seems to have corrected one or two typographical errors from the New York edition.

The text presented here (and that of Asa Gray's review) has been edited very lightly in order to make the writer's intentions clearer for readers today. The editorial changes include modernizing punctuation, standardizing the forms of citation as well as the use of ellipses and quotation marks, breaking up ungainly paragraphs, and amending minor matters of grammar. Hodge's system of capitalization, which is occasionally inconsistent, has been left intact, as have a few other inconsequential inconsistencies. In addition, the editors have augmented Hodge's subheads to indicate more clearly the thread of his argument.

All substantive material added by the editors in both the text and notes has been enclosed in brackets. Parentheses are from Hodge. Where quotations are not footnoted, Hodge did not footnote them. The editors' bracketed annotations translate foreign phrases, provide references for biblical quotations, briefly explain scientific and philosophical terms, identify the authorities Hodge cites, and highlight the major steps of his argument. To anticipate the companion volume to this work, a few of the notes refer to the views of Benjamin B. Warfield, Hodge's successor as professor of theology at Princeton Seminary. If an individual is not identified in the notes, pertinent biographical information was provided earlier in the work. Foreign phrases whose meaning is obvious (like *Gott und Natur*) are not translated. For biblical citations, Hodge mostly used the Authorized or King James Version, but occasionally seems to have provided his own translations.

The range of reading cited in this work testifies to Hodge's lifelong interest in scientific literature. The book is also a testimony to the reasoning powers of an active septuagenarian. At the same time, however, the form of *What Is Darwinism?* also reprises a genre that Hodge had practiced for nearly fifty years. Although it was published as a separate book, *What Is Darwinism?* is exactly the kind of essay-review that Hodge had regularly written for the *Princeton Review* throughout his long and influential career as editor of that journal. As so often in the *Princeton Review,* Hodge provides the most convincing case for his own views by setting them against the well-considered opinions of a worthy opponent. Rarely had there been an opponent as worthy as Charles Darwin. Admiring comments by the editors notwithstanding, it remains for readers to decide if Hodge was up to the task he took on himself.

Outline of Contents

Theories as to the Nature of the Universe
    The Scriptural Solution of the Problem
    The Pantheistic Theory
    Epicurean Theory
    Herbert Spencer's New Philosophy
    Hylozoic Theory
    Theism in Unscriptural Forms
Mr. Darwin's Theory
    Natural Selection
    Darwin's Use of the Word *Natural*
The Distinctive Element of Darwinism: Rejection of Teleology
    Darwin's Own Testimony
    Testimony of the Advocates of the Theory
        Alfred Russel Wallace
        Thomas Henry Huxley
        Ludwig Büchner
        Karl Vogt
        Ernst Haeckel
    Testimony of the Opponents of Darwinism
        The Duke of Argyll
        Louis Agassiz

What Is Darwinism?

Paul Janet
M. Flourens
Rev. Walter Mitchell
Principal Dawson
Relation of Darwinism to Religion
   Causes of the Alienation between Science and Religion
   Objections to Darwinism
   The Grand and Final Objection

What is Darwinism? This is a question which needs an answer. Great confusion and diversity of opinion prevail as to the real views of the man whose writings have agitated the whole world, scientific and religious. If a man says he is a Darwinian, many understand him to avow himself virtually an atheist, while another understands him as saying that he adopts some harmless form of the doctrine of evolution. This is a great evil.

It is obviously useless to discuss any theory until we are agreed as to what that theory is. The question, therefore, What is Darwinism? must take precedence of all discussion of its merits.

The great fact of experience is that the universe exists. The great problem which has ever pressed upon the human mind is to account for its existence. What was its origin? To what causes are the changes we witness around us to be referred? As we are a part of the universe, these questions concern ourselves. What are the origin, nature, and destiny of man? Professor Huxley[1] is right in saying,

> The question of questions for mankind—the problem which underlies all others, and is more interesting than any other—is the ascertainment of the place which Man occupies in nature and of his relation to the universe of things. Whence our race has come, what are the limits of our power over nature, and of nature's power over us, to what goal are we tending, are the problems which present themselves anew and with undiminished interest to every man born into the world.[2]

---

1. [Thomas Henry Huxley (1825–95), a zoologist and paleontologist, was one of Darwin's closest friends. Although he doubted whether Darwin's theory of natural selection adequately explained the transmutation of species, he so zealously defended *The Origin of Species* that he became known as "Darwin's bulldog."]

2. *Evidences of Man's Place in Nature.* London, 1864, p. 57.

Mr. Darwin undertakes to answer these questions. He proposes a solution of the problem which thus deeply concerns every living man. Darwinism is, therefore, a theory of the universe, at least so far as the living organisms on this earth are concerned. This being the case, it may be well to state, in few words, the other prevalent theories on this great subject so that the points of agreement and of difference between them and the views of Mr. Darwin may be the more clearly seen.

Theories as to the Nature of the Universe

The Scriptural Solution of the Problem

[The Scriptural] solution is stated in words equally simple and sublime: "In the beginning God created the heavens and the earth" [Gen. 1:1]. We have here, first, the idea of God. The word *God* has in the Bible a definite meaning. It does not stand for an abstraction, for mere force, for law or ordered sequence. God is a spirit, and as we are spirits we know from consciousness[3] that God is (1) A Substance; (2) That He is a person; and, therefore, a self-conscious, intelligent, voluntary agent. He can say I; we can address Him as Thou; we can speak of Him as He or Him. This idea of God pervades the Scriptures. It lies at the foundation of natural religion. It is involved in our religious consciousness. It enters essentially into our sense of moral obligation. It is inscribed ineffaceably, in letters more or less legible, on the heart of every human being. The man who is trying to be an atheist is trying to free himself from the laws of his being. He might as well try to free himself from liability to hunger or thirst.

The God of the Bible, then, is a Spirit, infinite, eternal, and unchangeable in his being, wisdom, power, holiness, goodness,

---

3. [Throughout this work Hodge appeals frequently to "consciousness" (as here and, e.g., p. 72), "the very constitution of our nature" (p. 66), "the testimony of the senses" (p. 131), "the intuitive perceptions and irresistible convictions of all mankind" (p. 153), and similar concepts. These notions were well-established themes in the commonsense philosophy that Hodge had learned at the College of New Jersey and at Princeton Seminary, and that he employed in most of his writings on the natural world and human society. For a discussion of the commonsense philosophy that formed such an important part of Hodge's training, see Mark A. Noll, *Princeton and the Republic, 1768–1822: The Search for a Christian Enlightenment in the Era of Samuel Stanhope Smith* (Princeton: Princeton University Press, 1989), 36–43, 188–94, and 290.]

and truth.[4] As every theory must begin with some postulate, this is the grand postulate with which the Bible begins. This is the first point.

The second point concerns the origin of the universe. It is not eternal either as to matter or form. It is not independent of God. It is not an evolution of his being or his existence form [i.e., form of existence]. He is extramundane [i.e., outside of the world] as well as antemundane [i.e., before the world]. The universe owes its existence to his will.

Thirdly, as to the nature of the universe, it is not a mere phenomenon. It is an entity, having real objective existence or actuality. This implies that matter is a substance endowed with certain properties in virtue of which it is capable of acting and of being acted upon. These properties, being uniform and constant, are physical laws to which, as their proximate causes, all the phenomena of nature are to be referred.

Fourthly, although God is extramundane, He is nevertheless everywhere present. That presence is not only a presence of essence, but also of knowledge and power. He upholds all things. He controls all physical causes, working through them, with them, and without them, as He sees fit. As we, in our limited spheres, can use physical causes to accomplish our purposes, so God everywhere and always cooperates with them to accomplish his infinitely wise and merciful designs.

Fifthly, man, a part of the universe, is, according to the Scriptures, as concerns his body, of the earth. So far, he belongs to the animal kingdom. As to his soul, he is a child of God, who is declared to be the Father of the spirits of all men. God is a spirit, and we are spirits. We are, therefore, of the same nature with God. We are Godlike, so that in knowing ourselves we know God. No man conscious of his manhood can be ignorant of his relationship to God as his Father.

The truth of this theory of the universe rests, in the first place, so far as it has been correctly stated, on the infallible authority of the Word of God. In the second place, it is a satisfactory solution of the problem to be solved—(1) It accounts for the origin of the universe. (2) It accounts for all the universe contains, and gives a satisfactory

---

4. [Hodge quotes here from the Westminster Shorter Catechism, question 4.]

explanation of the marvellous contrivances which abound in living organisms, of the adaptations of these organisms to conditions external to themselves, and of those provisions for the future which on any other assumption are utterly inexplicable. (3) It is in conflict with no truth of reason and with no fact of experience.[5] (4) The Scriptural doctrine accounts for the spiritual nature of man, and meets all his spiritual necessities. It gives him an object of adoration, love, and confidence. It reveals the Being on whom his indestructible sense of responsibility terminates. The truth of this doctrine, therefore, rests not only on the authority of the Scriptures, but on the very constitution of our nature. The Bible has little charity for those who reject it. It pronounces them to be either derationalized or demoralized, or both.

The Pantheistic Theory

[The Pantheistic theory] has been one of the most widely diffused and persistent forms of human thought on this whole subject. It has been for thousands of years not only the philosophy, but the religion of India, and, to a great extent, of China. It underlies all the forms of Greek philosophy. It crept into the Church, concealed under the disguise of Scriptural terminology, in the form of Neo-Platonism. It was constantly reappearing during the Middle Ages, sometimes in a philosophical, and sometimes a mystical form. It was

---

5. The two facts which are commonly urged as inconsistent with Theism are the existence of misery in the world, and the occurrence of undeveloped or useless organs, as teeth in the jaws of the whale and mammae on the breast of a man. As to the former objection, sin, which is the only real evil, is accounted for by the voluntary apostasy of man; and as to undeveloped organs they are regarded as evidences of the great plan of structure which can be traced in the different orders of animals. These unused organs were—says Professor Joseph Le Conte in his interesting volume on *Religion and Science,* New York, 1874, p. 54—regarded as blunders in nature, until it was discovered that use is not the only end of design. "By further patient study of nature," he says, "came the recognition of another law beside use—a law of order underlying and conditioning the law of use. Organisms are, indeed, contrived for use, but according to a preordained plan of structure, which must not be violated." It is of little moment whether this explanation be considered satisfactory or not. It would certainly be irrational to refuse to believe that the eye was made for the purpose of vision because we cannot tell why a man has mammae. A man might as well refuse to admit that there is any meaning in all the writings of Plato because there is a sentence in them which he cannot understand. [Joseph Le Conte (1823–1901), American geologist, was influenced by his teacher Louis Agassiz to see much design in nature. He came to identify himself as a "theistic evolutionist."]

revived by Spinoza in the seventeenth century, and subsequently became dominant in the philosophy and literature of Europe.[6] It is coming up again. Some distinguished naturalists are swinging round from one pole to the opposite—from saying there is no God to teaching that everything is God. Sometimes one and the same book in one half teaches materialism, in the other half idealism—the one affirming that everything is matter, the other that matter is nothing, but that everything is mind, and mind is God.

The leading principles of the Pantheistic theory are: (1) That there is an Infinite and Absolute Being. Of this Being nothing can be affirmed but actuality. It is denied that it is conscious, intelligent, or voluntary. (2) It is subject to the blind necessity of self-evolution or development. (3) This development being necessary is constant, from everlasting to everlasting. According to the Brahminical doctrine,[7] indeed, there are successive cycles of activity and repose, each cycle being measured by countless milliards [i.e., billions] of centuries. According to the moderns, self-evolution being necessary, there can be no repose, so that *Ohne Welt kein Gott* [without the world there is no God]. (4) The Finite is, therefore, the existence form of the Infinite; all that is in the latter for the time being is in the former. All that is possible is actual. (5) The Finite is the Infinite or, to use theistic language, the World is God in the sense that all the world is and contains is the form in which God at each successive moment exists. There is no power save only the power manifested in the world; [there is] no consciousness, intelligence, or voluntary activity but in finite things, and the aggregate of these is the power, consciousness, intelligence, and activity of God. What we call sin is as much a form of God's activity as what we call virtue. In other words, there is no such thing as free agency in man, no such thing as sin or responsibility. When a man dies he sinks into the abyss of being as a drop of water is lost in the ocean. (6) Man is the highest form of God's existence. God is incarnate in the human race.

---

6. [The Neo-Platonists believed all reality was derived from a single entity, variously labeled the One or the Good, which transcended all visible realities and remained essentially unknowable. Benedict de Spinoza (1632–77), Dutch lens-grinder and philosophical rationalist, was often regarded as Europe's most thoroughgoing pantheist. He proposed that all individual realities, whether bodies or ideas, were an extension of God's existence.]

7. [That is, of the Brahmins of India; Hodge's son, Archibald Alexander Hodge, had served briefly in India as a missionary.]

Strauss[8] says that what the Church teaches of Christ is not true of any individual man, but is true of mankind. Or, as Feuerbach[9] more concisely expresses it, "Man alone is our God." The blasphemy of some of the German philosophers on this subject is simply unutterable. In India we see the practical operation of this system when it takes hold on the people. There the personification of the Infinite as evil (the Goddess Kali) is the most popular object of worship.

Epicurean Theory

Epicurus[10] assumed the existence of matter, force and motion—*Stoff und Kraft* [matter and power]. He held that all space was filled with molecules of matter in a state of rapid motion in every direction. These molecules were subject to gravity and endowed with properties or forces. One combination of molecules gave rise to unorganized matter, another to life, another to mind; and from the various combinations, guided by unintelligent physical laws, all the wonderful organisms of plants and animals have arisen. To these combinations also all the phenomena of life, instinct, and intelligence in the world are to be referred.

This theory has been adopted in our day by a large class of scientific men, especially in Germany. The modern advocates of the theory are immeasurably superior to the ancient Epicureans in their knowledge of astronomy, botany, zoology, and biology; but in their theory of the universe, and in their mode of accounting for all the phenomena of life and intelligence, they are precisely on the same level. They have not added an idea to the system, which has ever been regarded as the opprobrium of human thought. Büchner, Moleschott, Vogt,[11] hold that matter is eternal and indestructible,

---

8. [David Friedrich Strauss (1808–74), a German historian of religion, held that most traditional ideas about Christianity were the product of humanity's capacity to fabricate myths, an idea expounded most clearly in his *Life of Jesus* (1835–36). Strauss figures later in this work as Hodge's premier representative of post-Christian atheism.]

9. [Ludwig Feuerbach (1804–72), German philosopher, exploited Georg Hegel's categories to argue that humans created the idea of God as a projection of their grandest notions about humanity itself.]

10. [Epicurus (341–270 B.C.) conceived of the physical world as an infinite supply of atoms moving in infinite space to form every possible variety of existence. John Tyndall (see p. 74, n. 19) identified him as a dominant figure in the history of the atomic philosophy.]

11. [Ludwig Büchner (1824–99) was considered the most influential German materialist of the nineteenth century. Jacob Moleschott (1822–93), who taught in

that matter and force are inseparable; the one cannot exist without the other. What, it is asked, is motion without something moving? What is electricity without an electrified body? What is attraction without molecules attracting each other? What is contractibility without muscular fibre, or secretion without a secreting gland? One combination of molecules exhibits the phenomena of life, another combination exhibits the phenomena of mind. All this was taught by the old heathen philosopher more than two thousand years ago. That this system denies the existence of God, of mind as a thinking substance distinct from matter, and of the possibility of the conscious existence of man after death, are not inferences drawn by opponents, but conclusions openly avowed by its advocates.

Herbert Spencer's New Philosophy

Mr. Darwin calls [Herbert] Spencer[12] our "great philosopher." His is the speculating mind of the new school of science. This gives to his opinions special interest, although no one but himself is to be held responsible for his peculiar views, except so far as others see fit to avow them. Mr. Spencer postulates neither mind nor matter. He begins with Force. Force, however, is itself perfectly inscrutable. All we know about it is that it is, that it is indestructible, and that it is persistent.

As to the origin of the universe, he says there are three possible suppositions: (1) That it is self-existent. (2) That it is self-created. (3) That it is created by an external agency.[13] All these he examines

---

Heidelberg, Zurich, and Rome, made scientific contributions in medicine and physiology while promoting his materialistic monism. Karl Vogt (1817–95), director of the Institute of Zoology in Geneva, eventually became a strong supporter of the Darwinian concept of natural selection.]

12. [Herbert Spencer (1820–1903), evolutionary theorist and author of a multivolume series on *Synthetic Philosophy*, the first volume of which was *First Principles*, never held a halfhearted view in his life. He was an extreme liberal individualist, an ardent proponent of social evolution, and a firm believer in the progress of Western society. Much of the later confusion in the discussion of Darwinism arose from the intermingling of Darwin's scientific conclusions and Spencer's grandiose claims for philosophic evolution. Spencer's own evolutionism, however, was closer to Jean Baptiste Lamarck's belief that changes acquired through environment could be passed on hereditarily (see p. 90, n. 36) than he was to Darwin's theory of natural selection.]

13. *First Principles of a New System of Philosophy*. By Herbert Spencer. Second edition. [New York, 1869, p. 30.]

and rejects. The first is equivalent to Atheism, by which Spencer understands the doctrine which makes Space, Matter, and Force eternal and the causes of all phenomena. This, he says, assumes the idea of self-existence, which is unthinkable. The second theory he makes equivalent to Pantheism. "The precipitation of vapor," he says, "into cloud, aids us in forming a symbolic conception of a self-evolved universe"; but, he adds, "really to conceive self-creation, is to conceive potential existence passing into actual existence by some inherent necessity, which we cannot do" (p. 32). The Theistic theory, he says, is equally untenable. "Whoever agrees that the atheistic hypothesis is untenable because it involves the impossible idea of self-existence, must perforce admit that the theistic hypothesis is untenable if it contains the same impossible idea" (p. 38). The origin of the universe is, therefore, a fact which cannot be explained. It must have had a cause, and all we know is that its cause is unknowable and inscrutable.

When we turn to nature, the result is the same. Everything is inscrutable. All we know is that there are certain appearances, and that where there is appearance there must be something that appears. But what that something is, what is the noumenon [Immanuel Kant's alternative for *thing-in-itself*, of which we have no direct knowledge] which underlies the phenomenon, it is impossible for us to know. In nature we find two orders of phenomena or appearances: the one objective or external, the other subjective in our consciousness. There are an Ego and a non-Ego, a subject and object. These are not identical. "It is," he says, "rigorously impossible to conceive that our knowledge is a knowledge of appearances only, without at the same time conceiving a reality of which they are appearances, for appearance without reality is unthinkable" (p. 88). So far we can go. There is a reality which is the cause of phenomena. Further than that, in that direction, our ignorance is profound. He proves that space cannot be an entity, an attribute, or a category of thought, or a nonentity. The same is true of time, of motion, of matter, of electricity, light, magnetism, etc., etc. They all resolve themselves into appearances produced by an unknown cause.

As the question, What is matter? is a crucial one, [Spencer] dwells upon it in various parts of his writings. Newton's theory of ultimate atoms, Leibnitz's doctrine of monads, and the dynamic theory of Boscovich, which makes matter mere centres of force, are all dismissed as "unthinkable."[14] It is not very clear in what sense that word is to be taken. Sometimes it seems to mean "meaningless"; at others, "self-contradictory or absurd"; at others, "inconceivable," i.e., that of which no conception or mental image can be formed; at any rate, it implies what is unknowable and untenable. The result is, so far as matter is concerned, that we know nothing about it. "Our conception of matter," he says, "reduced to its simplest shape, is that of coexistent positions that offer resistance, as contrasted with our conception of space in which the coexistent positions offer no resistance" (p. 166). Resistance, however, is a form of force; and, therefore, on the following page Spencer says "that forces standing in certain correlations, form the whole contents of our idea of matter."

When we turn from the objective to the subjective, from the external to the inward world, the result is still the same. [Spencer] agrees with Hume[15] in saying that the contents of our consciousness are a series of impressions and ideas. He dissents, however, from that philosopher in saying that that series is all we know. He admits that impressions necessarily imply that there is something that is impressed. He starts the question, What is it that thinks? and answers, We do not know (p. 63). He admits that the reality of individual personal minds, the conviction of personal existence is universal, and perhaps indestructible. Nevertheless that conviction cannot justify itself at the bar of reason; nay, reason is found to reject it (p. 65). Dean Mansel[16] says that consciousness gives us a knowledge of self as a substance and not merely of its varying states. This, however, [Spencer] says,

14. [Sir Isaac Newton (1642–1727) as mathematician and philosopher of the natural world was the symbol throughout the eighteenth and nineteenth centuries of the marvels that an unfettered pursuit of science could reveal. His firm belief in God was based on what he perceived as the extraordinary order of the universe. Gottfried Wilhelm von Leibniz (1646–1716), a native of Leipzig, was only slightly less famous than Newton as mathematician and philosopher of nature. Ruggiero Giuseppe Boscovich (or Boskovic, 1711–1787) propounded a theory of atoms in which the principle of energy is found in the atoms themselves.]

15. [David Hume (1711–76), notable Scottish skeptic, held that the conjunction of causes and effects in our minds is no guarantee of their causal connection in the world itself.]

16. [Henry Longueville Mansel (1820–71) was a High-Church Tory and dean of St. Paul's Cathedral. His Bampton Lectures of 1858 (*The Limits of Religious Thought*) borrowed from Immanuel Kant to propose a subjective apology for traditional Christianity.]

is absolutely negatived by the laws of thought. The fundamental condition to all consciousness, emphatically insisted upon by Mr. Mansel in common with Sir William Hamilton[17] and others, is the antithesis of subject and object.... What is the corollary from this doctrine, as bearing on the consciousness of self? The mental act in which self is known implies, like every other mental act, a perceiving subject and a perceived object. If, then, the object perceived is self, what is the subject that perceives? Or if it is the true self which thinks, what other self can it be that is thought of? Clearly, a true cognition of self implies a state in which the knowing and the known are one—in which subject and object are identified; and this Mr. Mansel rightly holds to be the annihilation of both. So that the personality of which each is conscious, and of which the existence is to each a fact beyond all others the most certain, is yet a thing which cannot be known at all; knowledge of it is forbidden by the very nature of human thought. (pp. 65–66)

Mr. Spencer does not seem to expect that any man will be shaken in his conviction by any such argument as that. When a man is conscious of pain, he is not to be puzzled by telling him that the pain is one thing (the object perceived) and the self another thing (the perceiving subject). He knows that the pain is a state of the self of which he is conscious. Consciousness is a form of knowledge, but knowledge of necessity supposes an intelligent reality which knows. A philosophy which cannot be received until men cease to believe in their own existence must be in extremis.

Mr. Spencer's conclusion is that the universe—nature, or the external world with all its marvels and perpetual changes—[and] the world of consciousness with its ever-varying states, are impressions or phenomena due to an inscrutable, persistent force.

As to the nature of this primal force or power, he quotes abundantly and approvingly from Sir William Hamilton and Mr. Mansel to prove that it is unknowable, inconceivable, unthinkable. He, however, differs from those distinguished writers in two points. While admitting that we know no more of the first cause than we do of a geometrical figure which is at once a circle and a square, yet we do know that it is actual. For this conviction we are not dependent on

---

17. [Sir William Hamilton (1788–1856), influential professor of philosophy at Edinburgh, attempted a reconciliation of Kantian subjective philosophy with the Scottish commonsense tradition to which Hodge adhered.]

faith. In the second place, Hamilton and Mansel taught that we know that the Infinite cannot be a person, self-conscious, intelligent, and voluntary; yet we are forced by our moral constitution to believe it to be an intelligent person. This Mr. Spencer denies. "Let those," he says, "who can, believe that there is eternal war between our intellectual faculties and our moral obligations. I, for one, admit of no such radical vice in the constitution of things" (p. 108). Religion has always erred, he asserts, in that while it teaches that the Infinite Being cannot be known, it insists on ascribing to it such and such attributes, which of course assumes that so far forth it is known. We have no right, he contends, to ascribe personality to the "Unknown Reality," or anything else, except that it is the cause of all that we perceive or experience. There may be a mode of being as much transcending intelligence and will as these transcend mechanical motion. To show the folly of referring to the Unknown the attributes of our own spirits, he makes "the grotesque supposition that the tickings and other movements of a watch constituted a kind of consciousness; and that a watch possessed of such a consciousness, insisted on regarding the watchmaker's actions as determined like its own by springs and escapements" (p. 111). The vast majority of men, instead of agreeing with Mr. Spencer in this matter, will doubtless heartily, each for himself, join the German philosopher Jacobi in saying, "I confess to Anthropomorphism inseparable from the conviction that man bears the image of God; and maintain that besides this Anthropomorphism, which has always been called Theism, is nothing but Atheism or Fetichism."[18]

Mr. Spencer, therefore, in accounting for the origin of the universe and all its phenomena, physical, vital, and mental, rejects Theism, or the doctrine of a personal God who is extramundane as well as antemundane, the creator and governor of all things; he rejects Pantheism, which makes the finite the existence form of the Infinite; he rejects Atheism, which he understands to be the doctrine of the eternity and self-existence of matter and force. He contents himself with saying we must acknowledge the reality of an unknown something which is the cause of all things—the noumenon of all

---

18. *Von den göttlichen Dingen* [*Of Divine Things*], *Werke* III, pp. 422, 425. Leipzig, 1816. [Friedrich Heinrich Jacobi (1743–1819) opposed what were thought to be the antireligious implications of Spinoza and Kant by defending feeling and faith as valid forms of theoretical knowledge.]

phenomena. "If science and religion are to be reconciled, the basis of the reconciliation must be this deepest, widest, and most certain of all facts—that the Power which the universe manifests is utterly inscrutable" (p. 46). "The ultimate of ultimates is Force." "Matter and motion, as we know them, are differently conditioned manifestations of Force." "If, to use an algebraic illustration, we represent Matter, Motion, and Force, by the symbols x, y, z; then we may ascertain the values of x and y in terms of z, but the value of z can never be found; z is the unknown quantity, which must forever remain unknown, for the obvious reason that there is nothing in which its value can be expressed" (pp. 169–70).

We have, then, no God but Force. "Atheist" is everywhere regarded as a term of reproach. Every man instinctively recoils from it. Even the philosophers of the time of the French Revolution repudiated the charge of atheism, because they believed in motion; and motion being inscrutable, they believed in an inscrutable something, i.e., in Force. We doubt not Mr. Spencer would indignantly reject the imputation of atheism; nevertheless, in the judgment of most men, the difference between Antitheist and Atheist is a mere matter of orthography.

Hylozoic Theory

[The hylozoic (life in matter)] theory assumes the universe to be eternal. There is nothing extra[mundane] or antemundane. There is but one substance, and that substance is matter. Matter, however, has an active and passive principle. Life and rationality are among its attributes or functions. The universe, therefore, is a living whole pervaded by a principle not only of life but of intelligence. This hylozoic doctrine some modern scientific men, as Professor Tyndall,[19] seem inclined to adopt. They tell us that matter is not the dead and degraded thing it is commonly regarded. It is active and transcendental. What that means we do not know. The word *transcendental* is like a parabola in that there is no knowing where its meaning ends. To say that matter is transcendental is saying there is no telling what it is up to. This habit of using words which have no definite meaning is very convenient to writers, but very much the reverse for readers.

19. [John Tyndall (1820–1893), prolific younger contemporary of Darwin, published widely on heat, crystals, zoology, physiology, and the relation of science and religion.]

Some of the ancient Stoics distinguished between the active and passive principles in the world, calling the one mind, the other matter. These however were as intimately united as matter and life in a plant or animal.

Theism in Unscriptural Forms

There are men who are constrained to admit the being of God, who depart from the Scriptural doctrine as to his relation to the world. According to some, God created matter and endowed it with certain properties, and then left it to itself to work out, without any interference or control on his part, all possible results. According to others, He created not only matter, but life, or living germs, one or more, from which without any divine intervention all living organisms have been developed. Others, again, refer not only matter and life, but mind also to the act of the Creator; but with creation his agency ceases. He has no more to do with the world than a shipbuilder has with the ship he has constructed, when it is launched and far off upon the ocean. According to all these views a creator is a mere *Deus ex machina* [extraneous mechanical contrivance introduced to solve an apparently unsolvable difficulty], an assumption to account for the origin of the universe.

Another general view of God's relation to the world goes to the opposite extreme. Instead of God doing nothing, He does everything.[20] Second causes have no efficiency. The laws of nature are said to be the uniform modes of divine operation. Gravitation does not flow from the nature of matter, but is a mode of God's uniform efficiency. What are called chemical affinities are not due to anything in different kinds of matter, but God always acts in one way in connection with an acid, and in another way in connection with an

20. [The view Hodge turns to in this paragraph comes very close to the general position advocated by three great Christian metaphysicians in the age of Newton: the French Catholic Nicolas de Malebranche (1638–1715), the Anglican bishop George Berkeley (1685–1753), and the Massachusetts Congregationalist Jonathan Edwards (1703–58). To these three, it was not only appropriate but necessary to see the world as an extension of God's purposive mind, since they thought that any other view, especially in an age fixated on materiality, would lead to the exclusion of God. Never feeling the threat of materialism as keenly as did Malebranche, Berkeley, and Edwards, the tradition of Scottish commonsense realism, to which Hodge subscribed, promoted a dualism between, on the one hand, the mind (or spirit) of God and of humanity and, on the other, the physical reality of the material world.]

alkali. If a man places a particle of salt or sugar on his tongue, the sensation which he experiences is not to be referred to the salt or sugar, but to God's agency. When this theory is extended, as it generally is by its advocates, from the external to the internal world, the universe of matter and mind, with all their phenomena, is a constant effect of the omnipresent activity of God. The minds of some men, as remarked above, are so constituted that they can pass from the theory that God does nothing to the doctrine that He does everything, without seeing the difference. Mr. Russel Wallace, the companion and peer of Mr. Darwin, devotes a large part of his book on *Natural Selection* to prove that the organs of plants and animals are formed by blind physical causes. Toward the close of the volume he teaches that there are no such causes. He asks the question, What is Matter? and answers, Nothing. We know, he says, nothing but force; and as the only force of which we have any immediate knowledge is mind-force, the inference is "that the whole universe is not merely dependent on, but actually *is*, the will of higher intelligences, or of one Supreme Intelligence."[21] This is a transition from virtual materialism to idealistic pantheism. The effect of this admission on the part of Mr. Wallace on the theory of natural selection is what an explosion of its boiler would be to a steamer in mid-ocean, which should blow out its deck, sides, and bottom. Nothing would remain above water.

The Duke of Argyll seems at times inclined to lapse into the same doctrine:

> Science in the modern doctrine of conservation of energy and the convertibility of forces, is already getting a firm hold of the idea, that all kinds of force are but forms of manifestations of one central force issuing from some one fountain-head of power. Sir John Herschel[22]

---

21. *The Theory of Natural Selection.* By Alfred Russel Wallace. London, 1870, p. 368. [Wallace (1823–1913) was exploring the Malay Archipelago when he wrote a scientific paper with conclusions that Darwin thought were the same as his "natural selection." That paper prompted Darwin to publish *The Origin of Species* and so make public ideas that had been germinating in his mind for nearly two decades. After several years of close cooperation with Darwin, Wallace became a spiritualist and argued that biological evolution is compatible with a religious view of humankind.]

22. [Sir John Frederick William Herschel (1792–1871), astronomer and onetime president of the British Association for the Advancement of Science, upheld the compatibility of rigorous scientific experimentation with relatively traditional religious beliefs.]

has not hesitated to say, "that it is but reasonable to regard the force of gravitation as the direct or indirect result of a consciousness or will existing somewhere." And even if we cannot certainly identify force in all its forms with the direct energies of the one Omnipresent and All-pervading Will, it is at least in the highest degree unphilosophical to assert the contrary—to think or to speak, as if the forces of nature were either independent of, or even separate from the Creator's power.[23]

The Duke, however, in the general tenor of his book, does not differ from the common doctrine, except in one point. He does not deny the efficiency of physical causes, or resolve them all into the efficiency of God; but he teaches that God, in this world at least, never acts except through those causes. He applies this doctrine even to miracles, which he regards as effects produced by second causes of which we are ignorant, that is, by some higher law of nature. The Scriptures, however, teach that God is not thus bound, that He operates through second causes, with them, or without them, as He sees fit. It is a purely arbitrary assumption that when Christ raised the dead, healed the lepers, or gave sight to the blind, any second cause intervened between the effect and the efficiency of his will. What physical law, or uniformly acting force, operated to make the axe float at the command of the prophet? [2 Kings 6:6] or, in that greatest of all miracles, the original creation of the world?[24]

Mr. Darwin's Theory

We have not forgotten Mr. Darwin. It seemed desirable, in order to understand his theory, to see its relation to other theories of the universe and its phenomena, with which it is more or less con-

23. *Reign of Law*. By the Duke of Argyle [*sic*]. Fifth edition, London, 1867, p. 123. [George Campbell, eighth Duke of Argyll (1823–1900), was a Scottish nobleman and British statesman who wrote several popular books on science and religion. They interpreted contemporary scientific findings within the framework of traditional religious beliefs.]

24. [In contending against his contemporaries' preoccupation with second causes for biblical miracles, Hodge set up an interesting contrast with B. B. Warfield, his most influential successor. One of Warfield's most distinguishing traits as a theological commentator on science was his reliance on a principle of *concursus* (whereby divine and natural causes may both be assigned to a single effect) to account for supernatural events in human history. He was, thus, less eager to divide natural from supernatural causes than Hodge is here.]

nected. His work on the *Origin of Species* does not purport to be philosophical. In this aspect it is very different from the cognate works of Mr. Spencer. Darwin does not speculate on the origin of the universe, on the nature of matter, or of force. He is simply a naturalist, a careful and laborious observer, skillful in his descriptions, and singularly candid in dealing with the difficulties in the way of his peculiar doctrine. He set before himself a single problem, namely, How are the fauna and flora of our earth to be accounted for? In the solution of this problem he assumes:

1. The existence of matter, although he says little on the subject. Its existence, however, as a real entity is everywhere taken for granted.

2. He assumes the efficiency of physical causes, showing no disposition to resolve them into mind-force, or into the efficiency of the First Cause.

3. He assumes also the existence of life in the form of one or more primordial germs. He does not adopt the theory of spontaneous generation. What life is he does not attempt to explain further than to quote (p. 326), with approbation, the definition of Herbert Spencer, who says, "Life depends on, or consists in, the incessant action and reaction of various forces"—which conveys no very definite idea.

4. To account for the existence of matter and life, Mr. Darwin admits a Creator. This is done explicitly and repeatedly. Nothing, however, is said of the nature of the Creator and of his relation to the world further than is implied in the meaning of the word.

5. From the primordial germ or germs (Mr. Darwin seems to have settled down to the assumption of only one primordial germ), all living organisms, vegetable and animal, including man, on our globe, through all the stages of its history, have descended.

6. As growth, organization, and reproduction are the functions of physical life, as soon as the primordial germ began to live, it began to grow, to fashion organs however simple for its nourishment and increase, and for the reproduction, in some way, of living forms like itself. How all living things on earth, including the endless variety of plants, and all the diversity of animals—insects, fishes, birds, the ichthyosaurus, the mastodon, the mammoth, and man—have descended from the primordial animalcule, he thinks, may be accounted for by the operation of the following natural laws:

First, the law of Heredity, or that by which like begets like. The offspring are like the parent.

Second, the law of Variation, that is, while the offspring are in all essential characteristics like their immediate progenitor, they nevertheless vary, more or less within narrow limits, from their parent and from each other. Some of these variations are indifferent, some deteriorations, some improvements, that is, they are such as enable the plant or animal to exercise its functions to greater advantage.

Third, the law of Over-Production. All plants and animals tend to increase in a geometrical ratio, and therefore tend to overrun enormously the means of support. If all the seeds of a plant, all the spawn of a fish, were to arrive at maturity, in a very short time the world could not contain them. Hence of necessity arises a struggle for life. Only a few of the myriads born can possibly live.

Fourth, here comes in the law of Natural Selection, or the Survival of the Fittest. That is, if any individual of a given species of plant or animal happens to have a slight deviation from the normal type, favorable to its success in the struggle for life, it will survive. This variation, by the law of heredity, will be transmitted to its offspring, and by them again to theirs. Soon these favored ones gain the ascendency, and the less favored perish; and the modification becomes established in the species. After a time another and another of such favorable variations occur, with like results. Thus very gradually, great changes of structure are introduced, and not only species, but genera, families, and orders in the vegetable and animal world are produced. Mr. Darwin says he can set no limit to the changes of structure, habits, instincts, and intelligence which these simple laws in the course of millions or milliards [i.e., billions] of centuries may bring into existence. He says, "We cannot comprehend what the figures 60,000,000 really imply, and during this, or perhaps a longer roll of years, the land and waters have everywhere teemed with living creatures, all exposed to the struggle for life, and undergoing change" (p. 354). "Mr. Croll,"[25] he tells us, "estimates that about

---

25. [James Croll (1821–1890) was a Christian of traditional views who worked as resident geologist in the Edinburgh office of the Geological Survey and who felt it was possible to construe Darwin's ideas in an orthodox way. The Cambrian period refers to the first segment of the Paleozoic era. The word was first used by Darwin's teacher, the Reverend Adam Sedgwick, to describe rock formations found in Wales (or "Cambria").]

sixty millions of years have elapsed since the Cambrian period, but this, judging from the small amount of organic change since the commencement of the glacial period, seems a very short time for the many and the great mutations of life, which have certainly occurred since the Cambrian formation; and the previous one hundred and forty million years can hardly be considered as sufficient for the development of the varied forms of life which certainly existed toward the close of the Cambrian period" (p. 379). Years in this connection have no meaning. We might as well try to give the distance of the fixed stars in inches. As astronomers are obliged to take the diameter of the earth's orbit as the unit of space, so Darwinians are obliged to take a geological cycle as their unit of duration.

Natural Selection

As natural selection which works so slowly is a main element in Mr. Darwin's theory, it is necessary to understand distinctly what he means by it. On this point he leaves us no room for doubt:

> This preservation of favorable variations, and the destruction of injurious variations, I call Natural Selection, or, the Survival of the Fittest.[26] (p. 92) . . .
>
> Owing to the struggle (for life) variations, however slight and from whatever cause proceeding, if they be in any degree profitable to the individuals of a species, in their infinitely complex relations to other organic beings and to their physical conditions of life, will tend to the preservation of such individuals, and will generally be inherited by their offspring. The offspring also will thus have a better chance of surviving, for, of the many individuals of any species which are periodically born, but a small number can survive. I have called this principle, by which each slight variation, if useful, is preserved, by the term Natural Selection, in order to mark its relation to man's power of selection. But the expression often used by Mr. Herbert Spencer of the Survival of the Fittest, is more accurate, and sometimes is equally convenient. (p. 72) . . .
>
> Slow though the progress of selection may be, if feeble man can do so much by artificial selection, I can see no limit to the amount of change, to the beauty and infinite complexity of the co-adaptations

---

26. ["Survival of the Fittest" was a phrase coined by Herbert Spencer. While Darwin used it, his own viewpoint would have been more accurately stated as "survival of the fit*ter*."]

between all organic beings, one with another, and with their physical conditions of life, which may be effected in the long course of time by nature's power of selection, or the survival of the fittest. (p. 125) . . .
It may be objected that if organic beings thus tend to rise in the scale, how is it that throughout the world a multitude of the lowest forms still exist; and how is it that in each great class some forms are far more highly developed than others? . . . On our theory the continuous existence of lowly forms offers no difficulty; for natural selection, or the survival of the fittest, does not necessarily include progressive development, it only takes advantage of such variations as arise and are beneficial to each creature under its complex relations of life. . . . Geology tells us that some of the lowest forms, the infusoria [e.g., protozoa] and rhizopods [e.g., amoebas] have remained for an enormous period in nearly their present state. (p. 145) . . .
The fact of little or no modification having been effected since the glacial period would be of some avail against those who believe in an innate and necessary law of development, but is powerless against the doctrine of natural selection, or the survival of the fittest, which implies only that variations or individual differences of a favorable nature occasionally arise in a few species and are then preserved. (p. 149)

This process of improvement under the law of natural selection includes not only chances[27] in the organic structure of animals, but also in their instincts and intelligence. On entering on this part of his subject, Mr. Darwin says, "I would premise that I have nothing to do with the origin of the primary mental powers, any more than I have with that of life itself. We are concerned only with the diversities of instinct and of other mental qualities within the same class" (p. 255). He shows that even in a state of nature the instincts of animals of the same species do in some degree vary, and that they are transmitted by inheritance. A mastiff has imparted courage to a greyhound, and a greyhound has transmitted to a shepherd-dog a disposition to hunt hares. Among sporting dogs, the young of the pointer or retriever have been known to point or to retrieve without instruction. "If," he says, "it can be shown that instincts do vary ever so little, then I can see no difficulty in natural selection preserving and continually accumulating variations of instinct to any extent

---

27. ["Chances" is almost certainly a misprint for "changes."]

that was profitable. It is thus, as I believe, that all the most complex and wonderful instincts have arisen" (p. 257).

[Mr. Darwin] was rather unguarded in saying that he saw no difficulty in accounting for the most wonderful instincts of animals. He admits that he has found very great difficulty. He selects three cases which he found it specially hard to deal with: that of the cuckoo, that of the cell-building bee, and of the slave-making ant. He devotes much space and labor in endeavoring to show how the instinct of the bee, for example, in the construction of its cell, *might* have been gradually acquired. It is clear, however, that he was not able fully to satisfy even his own mind; for he admits that "it will be thought that I have an overweening confidence in the principle of natural selection, when I do not admit that such wonderful and well established facts do not annihilate the theory" (p. 290). This remark was made with special reference to the instincts of the ant, which he finds very hard to account for. He adds,

> No doubt many instincts of very difficult explanation could be opposed to the theory of natural selection: cases in which we cannot see how an instinct could possibly have originated; cases in which no intermediate gradations are known to exist; cases of instinct of such trifling importance that they could hardly have been acted upon by natural selection; cases of instincts almost identically the same in animals so remote in the scale of nature that we cannot account for their similarity by inheritance from a common progenitor, and consequently cannot believe that they were independently acquired through natural selection. I will not here enter on those cases, but will confine myself to one special difficulty which at first appeared to me insuperable, and actually fatal to the whole theory. I allude to neuters, or sterile females in insect communities; for these neuters often differ widely in instinct and structure from both the males and the fertile females, and yet, from being sterile, they cannot propagate their kind. (p. 289)

He is candid enough to say, in conclusion, "I do not pretend that the facts given in this chapter (on instinct) strengthen in any great degree my theory; but none of the cases of difficulty, to the best of my judgment, annihilate it" (p. 297). When it is remembered that his theory is that slight variations occurring in an individual advantageous to it (not to its associates) in the struggle for life are perpetuated by inheritance, it is no wonder that the case of sterile ants gave him so much trouble. Accidental sterility is not favorable to the individual, and its being made permanent by inheritance is out of the question, for the sterile have no descendants. Yet these sterile females are not degenerations; they are in general larger and more robust than their associates.

We have thus seen that, according to Mr. Darwin, all the infinite variety of structure in plants and animals is due to the law of natural selection. "On the principle of natural selection with divergence of character," he says, "it does not seem incredible that, from some such low and intermediate form, both animals and plants have been developed, and if we admit this, we must likewise admit that all the organized beings which have ever lived on this earth may be descended from some one primordial form" (p. 573). We have seen also that he does not confine his theory to organic structure, but applies it to all the instincts and all the forms of intelligence manifested by irrational creatures. Nor does he stop there; he includes man within the sweep of the same law. "In the distant future I see open fields for far more important researches. Psychology will be based on a new foundation, that of the *necessary* acquirement of each mental power and capacity by gradation. Light will be thrown on the origin of man and his history" (p. 577).

The "distant future" was near at hand. In his introduction to his work on the *Descent of Man,* he says he had determined not to publish on that subject,

> as I thought that I should thus only add to the prejudices against my views. It seemed to me sufficient to indicate, in the first edition of my "Origin of Species," that by this work "light would be thrown on the origin of man and his history;" and this implies that man must be included with other organic beings in any general conclusion respecting his manner of appearance on this earth. Now the case wears a wholly different aspect. When a naturalist like Karl Vogt (we shall see in what follows what kind of a witness he is) ventures to say in his address as President of the National Institution of Geneva (1869), "Personne, en Europe au moins, n'ose plus soutenir la création indépendante et de toutes pièces, des espèces [There is no one, at least in Europe, who is so bold as to defend the independent and fully formed creation of every species]"—it is manifest that at least a large number of naturalists must admit that species are the modified descendants of other species; and this especially holds good of the younger and rising naturalists.... Of the older and honored chiefs

in natural science, many unfortunately are still opposed to evolution in every form.

Karl Vogt would not write thus. To him no man is honored who does not agree with him, and any man who believes in God he execrates.[28]

In 1871, Mr. Darwin ventured on the publication of his *Descent of Man*. In that work, he endeavors to show that the proximate progenitor of man is the ape. He says "there is less difference of structure between the two, than between the higher and lower forms of apes themselves." Not only so, but he attempts to show that the mental faculties of man are derived by slight variations, long continued, from the measure of intellect possessed by lower animals. He even says that there is less difference in intelligence between man and the higher mammals than there is between the intelligence of the ant and that of the coccus, insects of the same class.[29]

In like manner he teaches that man's moral nature has been evolved by slow degrees from the social instincts common to many animals (pp. 68, 94). The moral element, thus derived, he admits might lead to very different lines of conduct. "If men," he says, "were reared under the same conditions as hives-bees, there can hardly be a doubt, that our unmarried females would, like the worker-bees, think it a sacred duty to kill all their brothers, and mothers would strive to kill their fertile daughters; and no one would think of interfering" (vol. 1, p. 70).

"Lower animals, especially the dog, manifest love, reverence, fidelity, and obedience; and it is from these elements that the religious sentiment in man has been slowly evolved by a process of natural selection" (vol. 1, p. 65).

The grand conclusion is, "man (body, soul, and spirit) is descended from a hairy quadruped, furnished with a tail and pointed ears, probably arboreal in its habits, and an inhabitant of the Old World" (vol. 2, p. 372). Mr. Darwin adds: "He who denounces these views (as irreligious) is bound to explain why it is

---

28. [The New York edition of *What Is Darwinism?* lacked the "not" in this sentence, a typographical error corrected in the British publication. It is characteristic of Hodge, whose polemical battles in the *Princeton Review* were legion, to note Darwin's good grace in referring to his opponents, for Hodge was (almost always) a gracious intellectual combatant himself.]

29. *Descent of Man*, etc. By Charles Darwin, M.A., F.R.S., etc. [New York, 1871, vol. 1, p. 179.]

more irreligious to explain the origin of man as a distinct species by descent from some lower form, through the laws of variation and natural selection, than to explain the birth of the individual through the laws of ordinary reproduction" (vol. 2, p. 378).

Darwin's Use of the Word Natural

We have not yet reached the heart of Mr. Darwin's theory. The main idea of his system lies in the word *natural*. He uses that word in two senses: first, as antithetical to the word *artificial*. Men can produce very marked varieties as to structure and habits of animals. This is exemplified in the production of the different breeds of horses, cattle, sheep, and dogs, and specially, as Mr. Darwin seems to think, in the case of pigeons. Of these he says, "The diversity of breeds is something astonishing." Some have long, and some very short bills; some have large feet, some small; some long necks, others long wings and tails, while others have singularly short tails; some have thirty, and even forty, tail-feathers, instead of the normal number of twelve or fourteen. They differ as much in instinct as they do in form. Some are carriers, some pouters, some tumblers, some trumpeters; and yet all are descendants of the Rock Pigeon, which is still extant. If, then, he argues, man, in a comparatively short time, has by artificial selection produced all these varieties, what might be accomplished on the boundless scale of nature during the measureless ages of the geologic periods!

Secondly, he uses the word *natural* as antithetical to supernatural. Natural selection is a selection made by natural laws, working without intention and design. It is, therefore, opposed not only to artificial selection, which is made by the wisdom and skill of man to accomplish a given purpose, but also to supernatural selection, which means either a selection originally intended by a power higher than nature, or which is carried out by such power. In using the expression *Natural Selection,* Mr. Darwin intends to exclude design or final causes.[30] All the changes in structure, instinct, or

---

30. [Hodge begins his own argument at this point. Its success depends upon his demonstrating (a) that Darwin's distinctive teaching was not evolution per se or natural selection as a means of explaining the transmutation of species, but the exclusion of design from the development of species; (b) that Darwin really taught the exclusion of design; and (c) that such a view was a practical denial of the existence of God.]

intelligence in the plants or animals, including man, descended from the primordial germ or animalcule, have been brought about by unintelligent physical causes. On this point he leaves us in no doubt. He divines nature to be "the aggregate action and product of natural laws; and laws are the sequence of events as ascertained by us."

It [has] been objected that he often uses teleological language, speaking of purpose, intention, contrivance, adaptation, etc. In answer to this objection he says: "It has been said, that I speak of natural selection as a power or deity; but who objects to an author speaking of the attraction of gravity as ruling the movements of the planet?" He admits that in the literal sense of the words, natural selection is a false term; but "who ever objected to chemists, speaking of the elective affinities of various elements?—and yet an acid cannot strictly be said to elect the base with which it in preference combines" (p. 93). We have here an affirmation and a negation. It is affirmed that natural selection is the operation of natural laws, analogous to the action of gravitation and of chemical affinities. It is denied that it is a process originally designed or guided by intelligence, such as the activity which foresees an end and consciously selects and controls the means of its accomplishment. Artificial selection, then, is an intelligent process; natural selection is not.

There are in the animal and vegetable worlds innumerable instances of at least apparent contrivance which have excited the admiration of men in all ages. There are three ways of accounting for them. The first is the Scriptural doctrine, namely, that God is a Spirit, a personal, self-conscious, intelligent agent; that He is infinite, eternal, and unchangeable in his being and perfections; that He is ever present; that this presence is a presence of knowledge and power. In the external world there is always and everywhere indisputable evidence of the activity of two kinds of force: the one physical, the other mental. The physical belongs to matter and is due to the properties with which it has been endowed; the other is the everywhere-present and ever-acting mind of God. To the latter are to be referred all the manifestations of design in nature and the ordering of events in Providence. This doctrine does not ignore the efficiency of second causes; it simply asserts that God overrules and controls them. Thus the Psalmist says, "I am fearfully and wonderfully made. . . . My substance was not hid from thee, when I was made in secret, and curiously wrought (or embroidered) in the lower parts of the earth. Thine eyes did see my substance, yet being imperfect; and in thy book all my members were written, which in continuance were fashioned, when as yet there were none of them" [Ps. 139:14–16]. "He who fashioned the eye, shall not He see? He that formed the ear, shall not He hear?" [Ps. 94:9]. "God makes the grass to grow, and herbs for the children of men" [Ps. 104:14]. He sends rain, frost, and snow. He controls the winds and the waves. He determines the casting of the lot, the flight of an arrow, and the falling of a sparrow. This universal and constant control of God is not only one of the most patent and pervading doctrines of the Bible, but it is one of the fundamental principles of even natural religion.

The second method of accounting for contrivances in nature admits that they were foreseen and purposed by God, and that He endowed matter with forces which He foresaw and intended should produce such results. But here his agency stops. He never interferes to guide the operation of physical causes. He does nothing to control the course of nature or the events of history.[31] [With respect to] this theory it may be said (1) That it is utterly inconsistent with the Scriptures. (2) It does not meet the religious and moral necessities of our nature. It renders prayer irrational and inoperative. It makes it vain for a man in any emergency to look to God for help. (3) It is inconsistent with obvious facts. We see around us innumerable evidences of the constant activity of mind. This evidence of mind and of its operations, according to Lord Brougham and Dr. Whewell,[32] is far more clear than that of the existence of matter and of its forces. If one or the other is to be denied, it is the latter rather than the former. Paley[33] indeed says that if the construction of a watch be an

---

31. [Hodge's successor, B. B. Warfield, actually tried to combine what Hodge treats here as the first two distinct ways of accounting for contrivances in nature. For Warfield, it was possible to say both that God acted directly in "universal and constant control" of nature and that he "endowed matter with forces which He foresaw and intended should produce such results."]

32. [Henry Peter Brougham (1778–1868) was a British lawyer and Whig politician who published on miscellaneous scientific subjects, including *A Discourse on Natural Theology* (1835). William Whewell (1794–1866), Cambridge scientist and philosopher, was Britain's leading Christian metaphysician in the first half of the nineteenth century.]

33. [The natural theology of William Paley (1743–1805) established a benchmark for later theories. His *Natural Theology, or Evidences of the Existence and Attributes of the Deity* (1802) was especially renowned for arguing that the visible workings of

undeniable evidence of design, it would be a still more wonderful manifestation of skill if a watch could be made to produce other watches and, it may be added, not only other watches, but all kinds of time-pieces in endless variety. So it has been asked, if man can make a telescope, why cannot God make a telescope which produces others like itself? This is simply asking whether matter can be made to do the work of mind. The idea involves a contradiction. For a telescope to make a telescope supposes it to select copper and zinc in due proportions and fuse them into brass, to fashion that brass into inter-entering tubes, to collect and combine the requisite materials for the different kinds of glass needed, to melt them, grind, fashion, and polish them, [to] adjust their densities and focal distances, etc., etc. A man who can believe that brass can do all this might as well believe in God. The most credulous men in the world are unbelievers. The great Napoleon could not believe in Providence, but he believed in his star and in lucky and unlucky days.

This banishing God from the world is simply intolerable and, blessed be his name, impossible. An absent God who does nothing is, to us, no God. Christ brings God constantly near to us. He said to his disciples,

> Consider the ravens, for they neither sow nor reap; which have neither store-house nor barn; and God feedeth them; how much better are ye than the fowls? And which of you by taking thought can add to his stature one cubit? Consider the lilies how they grow; they toil not, neither do they spin; and yet I say unto you that Solomon in all his glory was not arrayed like one of these. If then God so clothe the grass, which is to-day in the field, and to-morrow is cast into the oven; how much more will He clothe you, O ye of little faith? And seek ye not what ye shall eat, or what ye shall drink, neither be ye of doubtful mind. For all these things do the nations of the world seek after; and your Father knoweth that ye have need of these things. [Matt. 6:26–32]

---

nature reflect the conscious activity of God. Darwin was trained to think about God and nature along lines dictated by Paley's arguments. One of the factors in Darwin's adult crisis of faith was his conclusion that the phenomena of nature do not so readily and directly show the beneficence of the Creator as Paley had suggested. On the Paleyite form of Darwin's reasoning even after he abandoned Paley's conclusions, see John H. Brooke, "The Relations between Darwin's Science and His Religion," in *Darwinism and Divinity*, ed. John Durant (New York: Basil Blackwell, 1985).]

It may be said that Christ did not teach science. True, but He taught truth; and science, so called, when it comes in conflict with truth, is what man is when he comes in conflict with God.

The advocates of these extreme opinions protest against being considered irreligious. Herbert Spencer says that his doctrine of an inscrutable, unintelligent, unknown force as the cause of all things is a much more religious doctrine than that of a personal, intelligent, and voluntary Being of infinite power and goodness. Matthew Arnold[34] holds that an unconscious "power which makes for right" is a higher idea of God than the Jehovah of the Bible. Christ says, God is a Spirit. Holbach[35] thought that he made a great advance on that definition when he said, God is motion.

The third method of accounting for the contrivances manifested in the organs of plants and animals is that which refers them to the blind operation of natural causes. They are not due to the continued cooperation and control of the divine mind, nor to the original purpose of God in the constitution of the universe. This is the doctrine of the Materialists, and to this doctrine, we are sorry to say, Mr. Darwin, although himself a theist, has given in his adhesion. It is on this account the Materialists almost deify him.

The Distinctive Element of Darwinism: Rejection of Teleology

From what has been said, it appears that Darwinism includes three distinct elements. First, evolution, or the assumption that all organic forms, vegetable and animal, have been evolved or developed from one, or a few, primordial living germs; second, that this evolution has been effected by natural selection, or the survival of the fittest; and third, and by far the most important and only distinctive element of his theory, that this natural selection is without design, being conducted by unintelligent physical causes. Neither the first nor the second of these elements constitutes Darwinism, nor do the two combined.

---

34. [Matthew Arnold (1822–88), poet and critic, thought an ethic embodying Victorian aesthetic ideals could adequately replace traditional Christian dogma as a guide to thought and life.]

35. [Baron Paul d'Holbach (1723–89), French *philosophe*, was notorious to nineteenth-century Christians as a promoter of philosophical materialism.]

As to the first, namely, evolution, Mr. Darwin himself, in the historical sketch prefixed to the fifth edition of his *Origin of Species*, says that Lamarck,[36] in 1811 and more fully in 1815, "taught that all species, including man, are descended from other species." He refers to some six or eight other scientists as teaching the same doctrine. This idea of Evolution was prominently presented and elaborated in the *Vestiges of Creation*, first published in 1844.[37] Ulrici, Professor in the University of Halle, Germany, in his work *Gott und die Natur* says that the doctrine of evolution took no hold on the minds of scientific men, but was positively rejected by the most eminent physiologists, among whom he mentions J. Müller, R. Wagner, Bischoff, Hoffmann, and others.[38]

The Rev. George Henslow, Lecturer on Botany at St. Bartholomew's Hospital, London, himself a pronounced evolutionist, says the theories of Lamarck and of the *Vestiges of Creation* have given place to that of Mr. Darwin,

> and there are not wanting many symptoms of decay in the acceptance even of his. Not only has he considerably modified his views in later editions of the "Origin of Species," distinctly expressing the opinion that he attributed too great influence to natural selection, but even men of science, Owen,[39] Huxley—and at least in its application to man, Wallace himself—are either opposed to it in great measure, or else give it but a qualified assent. Thus, it has been the fate of all the-

---

36. [Jean Baptiste Lamarck (1744–1829), French zoologist and natural philosopher who accounted for the diversity of plants and animals in terms of progressive evolution along predetermined lines as a result of environmental influence and the consequent use or disuse of certain organs. The inheritance of acquired characteristics is the only part of Lamarck's general theory of organic progress that remained of interest to later evolutionists.]

37. [*Vestiges of the Natural History of Creation* was the anonymously published work of the Edinburgh publisher Robert Chambers (1802–71). In an effective journalistic style Chambers popularized a doctrine of upward-moving, progressive evolution as demonstrated by the fossil series, which were the "vestiges" of his title.]

38. *Gott und die Natur.* By Dr. Hermann Ulrici. Second edition. Leipzig, 1866, p. 394. [As opponents of evolutionary ideas Ulrici (1806–84), who posed his theism against Hegelian pantheism, enlists Johannes Peter Müller (1801–58), Rudolf Wagner (1805–64), and Theodor Bischoff (1807–82) from his own era, and Friedrich Hoffmann (1660–1742) from an earlier day.]

39. [Richard Owen (1804–92) was the preeminent anatomist in Victorian England and, at first, a dedicated promoter of Darwin's scientific career. After the publication of *The Origin of Species*, however, Owen came to object to the theory of natural selection. (See also p. 50.)]

ories of the development of living things to lapse into oblivion. *Evolution* itself, however, will stand the same.[40]

We find in the *Transactions of the Victoria Institute* a still more decided repudiation of Darwinism on the part of Mr. Henslow. He there says: "I do not believe in Darwin's theory; and have endeavored to refute it by showing its utter impossibility."[41] He defines Evolution by saying,

> It supposes all animals and plants that exist now, or have ever existed, to have been produced through laws of generation from preexisting animals and plants respectively; that affinity amongst organic beings implies, or is due to community of descent; and that the degree of affinity between organisms is in proportion to their nearness of generation, or, at least, to the persistence of common characters, they being the products of originally the same parentage.[42]

A man, therefore, may be an evolutionist without being a Darwinian. It should be mentioned that Mr. Henslow expressly excludes man, both as to body and soul, from the law of evolution.

Nor is the theory of natural selection the vital principle of Mr. Darwin's theory, unless the word *natural* be taken in a sense antithetical to supernatural. In the historical sketch just referred to, Mr. Darwin not only says that he had been anticipated in teaching the doctrine of Evolution by Lamarck and the author of the *Vestiges of Creation,* but that the theory of natural selection, as the means of accounting for evolution, was not original with him. He tells us that as early as 1813 Dr. W. C. Wells "distinctly recognizes the principle of natural selection," and that Mr. Patrick Matthew in 1831 "gives precisely the same view of the origin of species as that propounded by Mr. Wallace and myself."[43] Ideas are like seed: they are often cast

---

40. *The Theory of Evolution of Living Things and the Application of the Principles of Evolution to Religion.* By Rev. George Henslow, M.A., F.L.S., F.G.S. London, 1873, pp. 27–28. [Henslow (1835–1925) held that scientific research revealed a purposeful direction in the development of species, and also that Paley's arguments about design in nature could be updated to fit into an evolutionary format.]

41. *Journal of the Transactions of the Victoria Institute, or Philosophical Society of Great Britain.* Vol. 4. London, 1870, p. 278.

42. *Evolution and Religion,* p. 29.

43. [William Charles Wells (1757–1817) had proposed a theory that those who are "fit" are most likely to survive in a given environment. In the appendix to *On*

forth, and not finding a congenial soil produce no fruit. To Mr. Darwin is undoubtedly due the elaboration and thoroughly scientific defence of the theory of natural selection, and to him is to be referred the deep and widespread interest which it has excited.

It is however neither evolution nor natural selection which gives Darwinism its peculiar character and importance. It is that Darwin rejects all teleology or the doctrine of final causes. He denies design in any of the organisms in the vegetable or animal world. He teaches that the eye was formed without any purpose of producing an organ of vision.

Although evidence on this point has already been adduced, yet as it is often overlooked, at least in this country, so that many men speak favorably of Mr. Darwin's theory, who are no more Darwinians than they are Mussulmans [i.e., Muslims]; and as it is this feature of his system which brings it into conflict not only with Christianity, but with the fundamental principles of natural religion, it should be clearly established. The sources of proof on this point are—(1) Mr. Darwin's own writings. (2) The expositions of his theory given by its advocates. (3) The character of the objections urged by its opponents.

The point to be proved is that it is the distinctive doctrine of Mr. Darwin that species owe their origin, not to the original intention of the divine mind, not to special acts of creation calling new forms into existence at certain epochs, not to the constant and everywhere operative efficiency of God, guiding physical causes in the production of intended effects, but to the gradual accumulation of unintended variations of structure and instinct, securing some advantage to their subjects.

Darwin's Own Testimony

That such is Mr. Darwin's doctrine we prove from his own writings. And the first proof from that source is found in express declarations. When an idea pervades a book and constitutes its character, detached passages constitute a very small part of the evidence of its being inculcated. In the present case, however, such passages are

---

Naval Timber and Arboriculture (1831), Patrick Matthew, a wealthy Scottish landowner, theorized about evolution by means of a process of natural selection. His enunciation of the theory had, however, no later impact.]

sufficient to satisfy even those who have not had occasion to read Mr. Darwin's books. In referring to the similarity of structure in animals of the same class, he says, "Nothing can be more hopeless than to attempt to explain this similarity of pattern in members of the same class, by utility or the doctrine of final causes."[44]

On the last page of his work, he says:

> It is interesting to contemplate a tangled bank, clothed with many plants of many kinds, with birds singing on the bushes, with various insects flitting about, and with worms crawling through the damp earth, and to reflect that these elaborately constructed forms, so different from each other, and dependent on each other in so complex a manner, have all been produced by laws acting around us. These laws, taken in the largest sense, being growth with reproduction; variability from the indirect and direct action of the conditions of life, and from use and disuse; a ratio of increase so high as to lead to a struggle for life, and as a consequence to natural selection, entailing divergence of character and extinction of less improved forms. Thus from the war of nature, from famine and death, the most exalted object which we are capable of conceiving, the production of the higher animals directly follows. There is a grandeur in this view of life, with its several powers, having been originally breathed by the Creator into a few forms or into one; and that whilst this planet has gone cycling on according to the fixed law of gravity, from so simple a beginning endless forms most beautiful and most wonderful have been, and are being evolved. (p. 579)

In another of his works, he asks,

> Did He (God) ordain that crop and tail-feathers of the pigeon should vary, in order that the fancier might make his grotesque pouter and fan-tail breeds? Did He cause the frame and mental qualities of the dog to vary, in order that a breed might be formed of indomitable ferocity, with jaws fitted to pin down the bull, for man's brutal sport? But if we give up the principle in one case; if we do not admit that the variations of the primeval dog were intentionally guided in order, for instance, that the greyhound, that perfect image of symmetry and vigor, might be formed; no shadow of reason can be assigned for the belief that variations, alike in nature and the results of the same general laws, which have been the groundwork through natural selection

---

44. *Origin of Species*, p. 517.

of the most perfectly adapted animals in the world, man included, were intentionally and specially guided. However much we may wish it, we can hardly follow Professor Asa Gray, in his belief "that variations have been led along certain beneficial lines, as a stream is led along useful lines of irrigation."[45]

Variations which by their gradual accumulation give rise to new species, genera, families, and orders are themselves, step by step, accidental. Mr. Darwin sometimes says they happen by chance; sometimes he says they happen of necessity; at others he says, "We are profoundly ignorant of their causes." These are only different ways of saying that they are not intentional. When a man lets anything fall from his hands and says it was accidental, he does not mean that it was causeless, he only means that it was not intentional. And that is precisely what Darwin means when he says that species arise out of accidental variations. His whole book is an argument against teleology. The whole question is, How are we to account for the innumerable varieties, kinds, and genera of plants and animals, including man? Were they intended, or did they arise from the gradual accumulations of unintentional variations? His answer to these questions is plain. On page 245 he says:

> Nothing at first can appear more difficult to believe than that the more complex organs and instincts have been perfected not by means superior to, though analogous with, human reason, but by innumerable slight variations, each good for the individual possessor. Nevertheless, this difficulty, though appearing to our imagination[46] insuperably great, cannot be considered real, if we admit the following propositions, namely, that all parts of the organizations and instincts offer, at least, individual differences; that there is a struggle for existence, which leads to the preservation of profitable deviations of structure or instinct; and, lastly, that gradations in the state of perfection of each organ may have existed, each good of its kind.

---

45. *The Variation of Animals and Plants under Domestication.* By Charles Darwin, F.R.S., etc. [New York, 1868, vol. 2, pp. 515–16. Asa Gray (1810–88), America's leading botanist of the nineteenth century and a longtime professor at Harvard, was among Darwin's earliest and most perceptive advocates in America. He was also a trinitarian Congregationalist of traditional theological views who felt that Darwin's conclusions fit into a teleological picture of the universe. His objections to Hodge's depiction of Darwin are found on pp. 160–69.]

46. What can the word *imagination* mean in this sentence, if it does not mean "common sense"?

He says, over and over, that if beauty or any variation of structure can be shown to be intended, it would "annihilate his theory." His doctrine is that such unintended variations, which happen to be useful in the struggle for life, are preserved on the principle of the survival of the fittest. He urges the usual objections to teleology derived from undeveloped or useless organs, as web-feet in the upland goose and frigate-bird, which never swim.

What, however, perhaps more than anything makes clear his rejection of design is the manner in which he deals with the complicated organs of plants and animals. Why don't he say, they are the product of the divine intelligence? If God made them, it makes no difference, so far as the question of design is concerned, how He made them: whether at once or by a process of evolution.[47] But instead of referring them to the purpose of God, he laboriously endeavors to prove that they may be accounted for without any design or purpose whatever.

"To suppose," he says, "that the eye with all its inimitable contrivances for adjusting the focus to different distances, for admitting different degrees of light, and for the correction of spherical and chromatic aberration, could have been formed by natural selection, seems, I freely confess, absurd in the highest degree" (p. 222). Nevertheless he attempts to explain the process:

> It is scarcely possible to avoid comparing the eye with the telescope. We know that this instrument has been perfected by the long continued efforts of the highest of human intellects; and we naturally infer that the eye has been formed by a somewhat analogous process. But may not this inference be presumptuous? Have we any right to assume that the Creator works by intellectual powers like those of man? If we must compare the eye to an optical instrument, we ought in imagination to take a thick layer of transparent tissue, with spaces filled with fluid, and with a nerve sensitive to light beneath, and then suppose every part of this layer to be continually changing slowly in density, so as to separate into layers of different densities and thicknesses, placed at different distances from each other, and with the surfaces of each layer slowly changing in form. Further, we must suppose that there is a power represented by natural selection, or the survival of the fittest, always intently watching each slight alteration in the

---

47. [In such an objection to Darwin, Hodge argues exactly as his successor, B. B. Warfield, would also later argue.]

transparent layers, and carefully preserving each, which, under varied circumstances, tends to produce a distinct image. We must suppose each new state of the instrument to be multiplied by the million; each to be preserved until a better is produced, and the old ones to be all destroyed. In living bodies, variations will cause the slight alterations, generation will multiply them almost infinitely, and natural selection will pick out with unerring skill each improvement.[48] (p. 226)

"Let this process," he says, "go on for millions of years," and we shall at last have a perfect eye.

It would be absurd to say anything disrespectful of such a man as Mr. Darwin, and scarcely less absurd to indulge in any mere extravagance of language; yet we are expressing our own experience when we say that we regard Mr. Darwin's books the best refutation of Mr. Darwin's theory. He constantly shuts us up to the alternative of believing that the eye is a work of design or the product of the unintended action of blind physical causes. To any ordinarily constituted mind, it is absolutely impossible to believe that it is not a work of design. Darwin himself, it is evident, dear as his theory is, can hardly believe it. "It is indispensable," he says, "to arrive at a just conclusion as to the formation of the eye, that the reason should conquer the imagination; but I have felt the difficulty far too keenly to be surprised at any degree of hesitation in extending the principle of natural selection to so startling an extent" (p. 225).

It will be observed that every step in his account of the formation of the eye is an arbitrary assumption. We must first assume a thick layer of tissue, then that the tissue is transparent, then that it has cavities filled with fluid, that beneath the tissue is a nerve sensitive to light, then that the fluid is constantly varying in density and thickness, that its surfaces are constantly changing their contour, that its different portions are ever shifting their relative distances, that every favorable change is seized upon and rendered permanent—thus after millions of years we may get an eye as perfect as that of an eagle.

48. Mr. Darwin's habit of personifying nature has given, as his friend Mr. Wallace says, his readers a good deal of trouble. He defines nature to be the aggregate of physical forces, and in the single passage quoted he speaks of Natural Selection as "intently watching," "picking out with unerring skill," and "carefully preserving." It is true, he tells us, that is all to be understood metaphorically. [Hodge's criticism in this note refers in passing to a problem in Darwin's presentation of his views that has been critiqued with considerable force in Robert M. Young, "Darwin's Metaphor: Does Nature Select?" *Monist* 55 (1971): 442–503.]

In like manner we may suppose a man to sit down to account for the origin and contents of the Bible, assuming as his "working hypothesis" that it is not the product of mind either human or divine, but that it was made by a typesetting machine worked by steam, and picking out type hap-hazard. In this way in a thousand years one sentence might be produced, in another thousand a second, and in ten thousand more, the two might get together in the right position. Thus in the course of "millions of years" the Bible might have been produced, with all its historical details, all its elevated truths, all its devout and sublime poetry, and above all with the delineation of the character of Christ, the ἰδέα τῶν ἰδεῶν [the Form of Forms], the ideal of majesty and loveliness, before which the whole world, believing and unbelieving, perforce bows down in reverence. And when reason has sufficiently subdued the imagination to admit all this, then by the same theory we may account for all the books in all languages in all the libraries in the world. Thus we should have Darwinism applied in the sphere of literature. This is the theory which we are told is to sweep away Christianity and the Church!

Mr. Darwin gives the same unsatisfactory account of the marvellous "contrivances" in the vegetable world. In one species of Orchids, the labellum or lower lip is hollowed into a great bucket continually filled with water secreted from two horns which stand above it; when the bucket is sufficiently filled, the water flows out through a pipe or spout on one side. The bees, which crowd into the flower for sake of the nectar, jostle each other, so that some fall into the water; and their wings becoming wet they are unable to fly, and are obliged to crawl through the spout. In doing this they come in contact with the pollen, which, adhering to their backs, is carried off to other flowers. This complicated contrivance by which the female plants are fertilized has, according to the theory, been brought about by the slow process of natural selection or survival of the fittest.

Still more wonderful is the arrangement in another species of orchids. When the bee begins to gnaw the labellum, he unavoidably touches a tapering projection which, when touched, transmits a vibration which ruptures a membrane which sets free a spring by which a mass of pollen is shot, with unerring aim, over the back of the bee, who then departs on his errand of fertilization.

A very large class of plants is fertilized by means of insects. These flowers are beautiful, not for the sake of beauty—for that Mr. Darwin says would annihilate his theory—but those which happen to be beautiful attract insects, and thus become fertilized and perpetuated, while the plainer ones are neglected and perish. So with regard to birds. The females are generally plain, because those of bright colors are so exposed during the period of incubation that they are destroyed by their enemies. In like manner male birds are usually adorned with brilliant plumage. This is accounted for on the ground that they are more attractive and thus they propagate their race, while the plainer ones have few or no descendants. Thus all design is studiously and laboriously excluded from every department of nature.

The preceding pages contain only a small part of the evidence furnished by Mr. Darwin's own writings that his doctrine involves the denial of all final causes. The whole drift of his books is to prove that all the organs of plants and animals, all their instincts and mental endowments, may be accounted for by the blind operation of natural causes, without any intention, purpose, or cooperation of God. This is what Professor Huxley and others call "the creative idea" to which the widespread influence of his writings is to be referred.

Testimony of the Advocates of the Theory

ALFRED RUSSEL WALLACE

It is time to turn to the exposition of Darwinism by its avowed advocates, in proof of the assertion that it excludes all teleology. The first of these witnesses is Mr. Alfred Russel Wallace, himself a distinguished naturalist. Mr. Darwin informs his readers that as early as 1844 he had collected his material and worked out his theory, but had not published it to the world, although it had been communicated to some of his friends. In 1858 he received a memoir from Mr. Wallace, who was then studying the natural history of the Malay Archipelago. From that memoir he learnt that Mr. Wallace had "arrived at almost exactly the same conclusions as I (he himself) have on the origin of species." This led to the publishing his book on that subject contemporaneously with Mr. Wallace's memoir. There has been no jealousy or rivalry between these gentlemen. Mr. Wallace gracefully acknowledges the priority of Mr. Darwin's claim, and attributes to him the credit of having elaborated and sustained it in a way to secure for it universal attention. These facts are mentioned in order to show the competency of Mr. Wallace as a witness as to the true character of Darwinism.

Mr. Wallace in *The Theory of Natural Selection* devotes a chapter to the consideration of the objections urged by the Duke of Argyll in his work on the *Reign of Law* against that theory. Those objections are principally two: first, that design necessarily implies an intelligent designer; and second, that beauty not being an advantage to its possessor in the struggle for life cannot be accounted for on the principle of the survival of the fittest. The Duke, he says, maintains that contrivance and beauty indicate "the constant supervision and interference of the Creator, and cannot possibly be explained by the unassisted action of any combination of laws. Now, Mr. Darwin's work," he adds, "has for its main object to show that all the phenomena of living things—all their wonderful organs and complicated structures, their infinite variety of form, size, and color, their intricate and involved relations to each other—may have been produced by the action of a few general laws of the simplest kind, laws which are in most cases mere statements of admitted facts" (p. 265). Those laws are those with which we are familiar: Heredity, Variations, Over-Production, Struggle for Life, Survival of the Fittest. "It is probable," he says,

> that these primary facts or laws are but results of the very nature of life, and of the essential properties of organized and unorganized matter. Mr. Herbert Spencer, in his "First Principles" and in his "Biology," has, I think, made us able to understand how this may be; but at present we may accept these simple laws, without going further back, and the question then is, Whether the variety, the harmony, the contrivance, and the beauty we perceive, can have been produced by the action of these laws alone, or whether we are required to believe in the incessant interference and direct action of the mind and will of the Creator.[49] (p. 267)

Mr. Wallace says that the Duke of Argyll maintains that God

---

[49]. The question is not, as Mr. Wallace says, "How has the Creator worked?" but it is, as he himself states, whether the essential properties of matter have alone worked out all the wonders of creation, or whether they are to be referred to the mind and will of God. It is worthy of remark how Messrs. Darwin and Wallace refer to Mr. Spencer as their philosopher. We have seen what Spencer's philosophy is.

has personally applied general laws to produce effects which those laws are not in themselves capable of producing; that the universe alone with all its laws intact, would be a sort of chaos, without variety, without harmony, without design, without beauty; that there is not (and therefore we may presume that there could not be) any self-developing power in the universe. I believe, on the contrary, that the universe is so constituted as to be self-regulating; that as long [as] it contains life, the forms under which that life is manifested have an inherent power of adjustment to each other and to their surroundings; and that this adjustment necessarily leads to the greatest amount of variety and beauty and enjoyment, because it does depend on general laws, and not on a continual supervision and rearrangement of details. . . .

The strange springs and traps and pitfalls found in the flowers of Orchids, cannot be necessary *per se*, since exactly the same end is gained in ten thousand other flowers which do not possess them. Is it not then an extraordinary idea, to imagine the Creator of the universe contriving the various complicated parts of these flowers, as a mechanic might contrive an ingenious toy or a difficult puzzle? Is it not a more worthy conception, that they are the results of those general laws which were so coordinated at the first introduction of life upon the earth as to result necessarily in the utmost possible development of varied forms? . . .

I for one cannot believe that the world would come to chaos if left to law alone. . . . If any modification of structure could be the result of law, why not all? If some self-adaptations should arise, why not others? If any varieties of color, why not all the varieties we see? No attempt is made to explain this except by reference to the fact that "purpose" and "contrivance" are everywhere visible, and by an illogical deduction they could only have arisen by the direct action of some mind, because the direct action of our minds produces similar "contrivances"; but it is forgotten that adaptation, however produced, must have the appearance of design.[50] (pp. 268, 270, 280)

After referring to the fact that florists and breeders can produce varieties in plants and animals so that "whether they wanted a bulldog to torture another animal, a greyhound to catch a hare, or a bloodhound to hunt down their oppressed fellow-creatures, the required variations have always appeared," he adds:

50. It is, therefore, clear that design is what Mr. Darwin and Mr. Wallace repudiate.

To be consistent, our opponents must maintain that every one of the variations that have rendered possible the changes produced by man, have been determined at the right time and place by the Creator. Every race produced by the florist or breeder, the dog or the pigeon fancier, the rat-catcher, the sporting man, or the slave-hunter, must have been provided for by varieties occurring when wanted; and as these variations were never withheld, it would prove that the sanction of an all-wise and all-powerful Being has been given to that which the highest human minds consider to be trivial, mean, or debasing.[51] (p. 290)

The Nebular Hypothesis, as propounded by La Place,[52] proposed to account for the origin of the universe by a process of evolution under the control of mere physical forces. That hypothesis has, so far as evolution is concerned, been adopted by men who sincerely believe in God and in the Bible. But they hold not only that God created matter and endowed it with its properties, but that He designed the universe and so controlled the operation of physical laws that they accomplished his purpose. So there are Christian men who believe in the evolution of one kind of plants and animals out of earlier and simpler forms; but they believe that everything was designed by God, and that it is due to his purpose and power that all the forms of vegetable and animal life are what they are.[53] But this is not the question. What Darwin and the advocates of his theory deny is all design. The organs, even the most complicated and wonderful, were not intended. They are said to be due to the undirected and unintended operation of physical laws. This is Mr. Wallace's argument. He endeavors to show that it is unworthy of God that He should be supposed to have contrived the mechanism of the orchids, as a mechanist contrives a curious puzzle.

51. That God permits men in the use of the laws of nature to distil alcohol and brew poisons does not prove that He approves of drunkenness or murder.
52. [Pierre Simon de Laplace (1749–1827), French astronomer and cosmologist, proposed his nebular hypothesis in 1796: the origin of the solar system was a natural development over extended periods of time.]
53. [Among those holding this belief were America's leading evolutionist, Asa Gray; George Frederick Wright, a Congregational minister who eventually cooperated with Gray in publishing works on the reconciliation of natural selection and historic Protestant orthodoxy; and, closer to Hodge at Princeton, his theological successor, B. B. Warfield, and James McCosh, a Presbyterian minister who was president of the College of New Jersey (which Hodge served as trustee).]

We recently heard Prof. Joseph Henry,[54] in a brief address, say substantially: "If I take brass, glass, and other materials, and fuse them, the product is a slag. This is what physical laws do. If I take those same materials, and form them into a telescope, that is what mind does." This is the whole question in a nutshell. That design implies an intelligent designer is a self-evident truth. Every man believes it, and no man can practically disbelieve it. Even those naturalists who theoretically deny it, if they find in a cave so simple a thing as a flint arrow-head, are as sure that it was made by a man as they are of their own existence. And yet they want us to believe that an eagle's eye is the product of blind natural causes. No combination of physical forces ever made a ship or a locomotive. It may, indeed, be said that they are dead matter, whereas plants and animals live. But what is life but one form of the organizing efficiency of God?

Mr. Wallace does not go as far as Mr. Darwin. He recoils from regarding man either as to body or soul as the product of mere natural causes. He insists that "a superior intelligence is necessary to account for man" (p. 359). This of course implies that the agency of no such higher intelligence is admitted in the production of plants or of animals lower than man.

Thomas Henry Huxley

The second witness as to the character of Mr. Darwin's theory is Professor Huxley. We have some hesitation in including the name of this distinguished naturalist among the advocates of Darwinism.[55]

---

54. [Joseph Henry (1797–1878) was Hodge's friend and a professor of natural science at the College of New Jersey. Henry later became the founding secretary of the Smithsonian Institution and a great supporter of general scientific research in America.]

55. Mr. Huxley, if we may judge from what he says of himself, is somewhat liable to be misunderstood. He says he was fourteen years laboring to resist the charge of Positivism made against the class of scientific men to which he belongs. He also tells us in his letter to Professor Tyndall, prefixed to his volume of *Lay Sermons and Addresses*, that the "Essay on the Physical Basis of Life" included in that volume was intended as a protest, from the philosophical side, against what is commonly called Materialism. It turned out, however, that the public regarded it as an argument in favor of Materialism. This we think was a very natural, if not an unavoidable mistake on the part of the public. For in that essay he says that Protoplasm, or the physical basis of life, "is a kind of matter common to all living beings, that the powers or faculties of all kinds of living matter, diverse as they be in degree, are substantially of the same kind." Protoplasm as far as examined contains the four elements—carbon,

The Distinctive Element of Darwinism: Rejection of Teleology

On the one hand, in his essay on the *Origin of Species,* printed in the *Westminster Review* in 1860, and reprinted in his *Lay Sermons* in 1870, he says:

> There is no fault to be found with Mr. Darwin's method, but it is another thing whether he has fulfilled all the conditions imposed by that method. Is it satisfactorily proved that species may[56] be originated by selection? that none of the phenomena exhibited by species are inconsistent with the origin of species in this way? If these questions can be answered in the affirmative, Mr. Darwin's view steps out of the rank of hypotheses into that of theories;[57] but so long as the evidence at present adduced falls short of enforcing that affirmative, so long, to our minds, the new doctrine must be content to remain

---

hydrogen, oxygen, and nitrogen. These are lifeless bodies, "but when brought together under certain conditions, they give rise to the still more complex body Protoplasm; and this protoplasm exhibits the phenomena of life." There is no more reason, he teaches, for assuming the existence of a mysterious something called vitality to account for vital phenomena than there is for the assumption of something called Aquasity to account for the phenomena of water. Life is said to be "the product of a certain disposition of material molecules." The matter of life is "composed of ordinary matter, differing from it only in the manner in which its atoms are aggregated. I take it," he says, "to be demonstrable that it is utterly impossible to prove that anything whatever may not be the effect of a material and necessary cause, and that human logic is equally incompetent to prove that any act is really spontaneous. A really spontaneous act is one, which, by the assumption, has no cause; and the attempt to prove such a negative as this, is on the face of the matter absurd. And while it is thus a philosophical impossibility to demonstrate that any given phenomenon is not the effect of a material cause, any one who is acquainted with the history of science will admit that its progress has, in all ages, meant, and now more than ever means, the extension of what we call matter and causation, and the concomitant gradual banishment from all regions of human thought of what we call spirit and spontaneity."

56. It cannot escape the attention of any one that Mr. Darwin, Mr. Wallace, Professor Huxley, and all the other advocates or defenders of Darwinism, do not pretend to prove anything more than that species *may* be originated by selection, not that there is no other satisfactory account of their origin. Mr. Darwin admits that referring them to the intention and efficiency of God accounts for everything, but, he says, that is not science.

57. [In the categories of the commonsense philosophy to which Hodge subscribed, a strong distinction was made between "theories," which arose inductively from data collected by the physical or moral senses, and "hypotheses," which were regularly derided as whimsical philosophical speculations unrelated to the domain of ascertainable facts. In this Hodge and similar thinkers paid homage to Newton's claim, "I frame no hypotheses" (*non fingo hypotheses*).]

among the former—an extremely valuable, and in the highest degree probable, doctrine; indeed, the only extant hypothesis which is worth anything in a scientific point of view; but still a hypothesis, and not yet a theory of species. After much consideration and assuredly with no bias against Mr. Darwin's views, it is our clear conviction that, as the evidence now stands, it is not absolutely proven that a group of animals, having all the characters exhibited by species in Nature, has ever been originated by selection, whether artificial or natural.[58]

Again, in his work on *Man's Place in Nature,* Huxley expresses himself much to the same effect:

A true physical cause is admitted to be such only on one condition, that it shall account for all the phenomena which come within the range of its operation. If it is inconsistent with any one phenomenon it must be rejected; if it fails to explain any one phenomenon it is so far to be suspected, though it may have a perfect right to provisional acceptance.... Our acceptance, therefore, of the Darwinian hypothesis must be provisional so long as one link in the chain of evidence is wanting; and so long as all the animals and plants certainly produced by selective breeding from a common stock are fertile, and their progeny are fertile one with another, that link will be wanting. For so long selective breeding will not be proved to be competent to all that is required if it produce natural species.[59]

(In immediate connection with the above passage, there is another which throws a clear light on Professor Huxley's cosmical views:

The whole analogy of natural operations furnish so complete and crushing an argument against the intervention of any but what are called secondary causes, in the production of all the phenomena of the universe; that, in view of the intimate relations of man and the rest of the living world, and between the forces exerted by the latter and all other forces, I can see no reason for doubting that all are coordinate terms of nature's great progression, from formless to

58. *Lay Sermons, Addresses, and Reviews.* By Thomas Henry Huxley, LL.D., F.R.S. London, 1870, p. 323.
59. *Evidence of Man's Place in Nature.* London, 1864, p. 107.

formed, from the inorganic to the organic, from blind force to conscious intellect and will.[60])

Ought not this to settle the matter? Are we to give up the Bible and all our hopes for the sake of an hypothesis that all living things, including man, on the face of the earth are descended from a primordial animalcule by natural selection, when such a man as Huxley, who (as Voltaire said of the prophet Habakkuk) is *capable de tout* [liable to say anything], says that it has not been proved that any one species has thus originated?

But on the other hand, while he honestly admits that Darwin's doctrine is a mere hypothesis and not a theory, he has nevertheless written at least three essays or reviews in its exposition and vindication. He is freely referred to on the continent of Europe, at least, as an ardent advocate of the doctrine; and he quotes without protest such designations of himself. At any rate, as he assures his readers that he has no bias against Mr. Darwin's views, as he has devoted much time and attention to the subject, and as he is one of the most prominent naturalists of the age, there can be no question as to his competency as a witness as to what Darwinism is.

His testimony that Mr. Darwin's doctrine excludes all teleology or final causes is explicit. In his review of the "Criticisms on the *Origin of Species*" he says

> that when he first read Mr. Darwin's book, that which struck him most forcibly was the conviction that teleology, as commonly understood, had received its death-blow at Mr. Darwin's hands. For the teleological argument runs thus: An organ is precisely fitted to perform a function or purpose; therefore, it was specially constructed to per-

---

60. Since writing the above paragraph our eye fell on the following note on the 89th page of the Duke of Argyle's [*sic*] *Reign of Law*, which it gives us pleasure to quote. It seems that a writer in the *Spectator* had charged Professor Huxley with Atheism. In the number of that paper for February 10, 1866, the Professor replies: "I do not know that I care very much about popular odium, so there is no great merit in saying that if I really saw fit to deny the existence of a God I should certainly do so, for the sake of my own intellectual freedom, and be the honest atheist you are pleased to say I am. As it happens, however, I cannot take this position with honesty, inasmuch as it is, and always has been, a favorite tenet, that Atheism is as absurd, logically speaking, as Polytheism." In the same paper he says, "The denying the possibility of miracles seems to me quite as unjustifiable as speculative Atheism." How this can be reconciled with the passages quoted above, we are unable to see.

form that function. In Paley's famous illustration, the adaptation of all the parts of a watch to the function or purpose of showing the time, is held to be evidence that the watch was specially contrived to that end; on the ground that the only cause we know of competent to produce such an effect as a watch which shall keep time, is a contriving intelligence adapting the means directly to that end.[61]

This, Mr. Huxley tells us, is precisely what Darwin denies with reference to the organs of plants and animals. The eye was not formed for the purpose of seeing, or the ear for hearing. It so happened that a nerve became sensitive to light; then in course of time it happened that a transparent tissue came over it, and thus in "millions of years" an eye, as we have seen above, happened to be formed. No such organ was ever intended or designed by God or man. "An apparatus," says Professor Huxley,

> thoroughly adapted to a particular purpose, might be the result of a method of trial and error worked by unintelligent agents, as well as by the application of means appropriate to the end by an intelligent agent. . . . For the notion that every organism has been created as it is and launched straight at a purpose, Mr. Darwin substitutes the conception of something, which may fairly be termed a method of trial and error. Organisms vary incessantly; of these variations the few meet with surrounding conditions which suit them, and thrive; the many are unsuited, and become extinguished. . . . For the teleologist an organism exists, because it was made for the conditions in which it is found; for the Darwinian an organism exists, because, out of many of its kind, it is the only one which has been able to persist in the conditions in which it is found. . . . If we apprehend the spirit of the "Origin of Species" rightly, then, nothing can be more entirely and absolutely opposed to teleology, as it is commonly understood, than the Darwinian theory. (pp. 302–3)

It has already been stated that Mr. Wallace does not apply the doctrine of evolution to man; neither does Mr. Mivart,[62] a distinguished naturalist, who is a member of the Latin [i.e., Roman Cath-

---

61. *Lay Sermons*, p. 330.
62. [St. George Mivart (1827–1900) was a Roman Catholic evolutionist whom Darwin considered perhaps his sharpest critic. Mivart held both that natural selection cannot explain anomalies in the natural world and that evolution on the whole is compatible with traditional Christianity. For his pains he was ostracized by the scientific community and excommunicated from the Catholic Church.]

olic] Church. The manner in which Professor Huxley speaks of these gentlemen shows how thoroughly, in his judgment, Mr. Darwin banishes God from his works:

> Mr. Wallace and Mr. Mivart are as stout evolutionists as Mr. Darwin himself; but Mr. Wallace denies that man can have been evolved from a lower animal by that process of natural selection, which he, with Mr. Darwin, holds to be sufficient for the evolution of all animals below man; while Mr. Mivart, admitting that natural selection has been one of the conditions of the animals below man, maintains that natural selection must, even in their case, have been supplemented by some other cause—of the nature of which, unfortunately, he does not give us any idea. Thus Mr. Mivart is less of a Darwinian than Mr. Wallace, [who] has faith in the power of natural selection. But he is more of an evolutionist than Mr. Wallace, because Mr. Wallace thinks it necessary to call in an intelligent agent, a sort of supernatural Sir John Sebright, to produce even the animal frame of man; while Mr. Mivart requires no Divine assistance till he comes to man's soul.[63]

63. [Sir John Sebright (1767–1846) was a politician and agriculturalist best known for his theories on the breeding of domestic animals. Hodge's own note at this point is as follows:] *Contemporary Review*, vol. 18, 1871, p. 444. In this same article Mr. Huxley says: "Elijah's great question, Will ye serve God or Baal? Choose ye, is uttered audibly enough in the ears of every one of us as we come to manhood. Let every man who tries to answer it seriously ask himself whether he can be satisfied with the Baal of authority, and with all the good things his worshippers are promised in this world and the next. If he can, let him, if he be so inclined, amuse himself with such scientific implements as authority tells him are safe and will not cut his fingers; but let him not imagine that he is, or can be, both a true son of the Church and a loyal soldier of science.... And, on the other hand, if the blind acceptance of authority appear to him in its true colors, as mere private judgment *in excelsis*, and if he have courage to stand alone face to face with the abyss of the Eternal and Unknowable, let him be content, once for all, not only to renounce the good things promised by 'Infallibility,' but even to bear the bad things which it prophesies; content to follow reason and fact in singleness and honesty of purpose, wherever they may lead, in the sure faith that a hell of honest men will to him be more endurable than a paradise full of angelic shams." There can be no doubt that the Apostle Paul believed in the infallibility of the Scriptures. Imagine Professor Huxley calling St. Paul to his face a sham! What are all the Huxleys who have ever lived, or ever can live, to that one Paul in power for good over human thought, character, and destiny?

Professor Huxley goes on in the next paragraph to say: "Mr. Mivart asserts that 'without belief in a personal God there is no religion worthy of the name.' This is a matter of opinion. But it may be asserted, with less reason to fear contradiction, that the worship of a personal God, who, on Mr. Mivart's hypothesis, must have used words studiously calculated to deceive his creatures and worshippers, is 'no religion worthy of the name.' '*Incredible est, Deum illis verbis ad populum fuisse locutum quibus de-*

WHAT IS DARWINISM?

In the *Academy* for October, 1869, there is a review by Professor Huxley of Dr. Haeckel's[64] *Natürliche Schöpfungsgeschichte* [*Natural History of Creation*], in which he says:

> Professor Haeckel enlarges on the service which the "Origin of Species" has done in favoring what he terms "the causal or mechanical" view of living nature as opposed to the "teleological or vitalistic" view. And no doubt it is quite true the doctrine of evolution is the most formidable opponent of all the commoner and coarser forms of teleology. Perhaps the most remarkable service to the philosophy of Biology rendered by Mr. Darwin is the reconciliation of Teleology and Morphology, and the explanation of the facts of both which his view offers.
>
> The teleology which supposes that the eye, such as we see it in man or in the higher vertebrata, was made with the precise structure which it exhibits, to make the animal which possesses it to see, has undoubtedly received its death-blow. But it is necessary to remember that there is a higher teleology, which is not touched by the doctrine of evolution, but is actually based on the fundamental proposition of evolution. That proposition is, that the whole world, living and not living, is the result of the mutual interaction, according to definite laws, of forces possessed by the molecules of which the primitive nebulosity of the universe was composed. If this be true, it is no less certain that the existing world lay potentially in the cosmic vapor; and that a sufficient intelligence could, from a knowledge of the properties of that vapor, have predicted, say, the state of fauna of Great Britain in 1869, with as much certainty as one can say what will happen to the vapor of the breath on a cold winter's day.

This is the doctrine of the self-evolution of the universe. We know not what may lie behind this in Mr. Huxley's mind, but we are very sure that there is not an idea in the above paragraph which Epicurus of old, and Büchner, Vogt, Haeckel, and other *Materialisten von Pro-*

---

ciperetur [It is incredible that God spoke with those words to the people in order to deceive them]' is a verdict in which for once Jesuit casuistry concurs with the healthy moral sense of all mankind" (p. 458). Mr. Huxley calls believers in the Scriptures, and (apparently) believers in a personal God, bigots, old ladies of both sexes, bibliolaters, fools, etc., etc.

64. [Ernst Haeckel (1834–1919) was such an enthusiastic promoter of Darwinism that Darwin himself felt ill at ease as Haeckel expanded what he considered the key insights of *The Origin of Species* into a comprehensive approach to the natural world. For all his adulation of Darwin, however, Haeckel's version of evolution was less Darwinian than Neo-Lamarckian.]

fession [professional materialists] would not cheerfully adopt. His distinction between a higher and lower teleology is of no account in this discussion. What is the teleology to which, he says, Mr. Darwin has given the death-blow, the extracts given above clearly show. The eye, Huxley says, was not made for the purpose of seeing, or the ear for the purpose of hearing. "According to teleology," he says, "each organism is like a rifle bullet fired straight at a mark; according to Darwin, organisms are like grapeshot, of which one hits something and the rest fall wide."[65]

LUDWIG BÜCHNER

Dr. Ludwig Büchner, president of the medical association of Hessen-Darmstadt, etc., is not only a man of science but a popular writer. Perhaps no book of its class, in our day, has been so widely circulated as his volume on *Kraft und Stoff* (*Force and Matter*). It has been translated into all the languages of Europe. He holds that matter and force are inseparable; there cannot be the one without the other; both are eternal and imperishable; neither can be either increased or diminished; life originated spontaneously by the combination of molecules of matter under favorable conditions; all the phenomena of the universe, inorganic and organic, whether physical, vital, or mental, are due to matter and its forces. Consequently there is no God, no creation, no mind distinct from matter, no conscious existence of man after death. All this is asserted in the most explicit terms. Dr. Büchner has published a work on Darwinism in two volumes. Darwin's theory, he says, "is the most thoroughly naturalistic that can be imagined, and far more atheistic than that of his decried predecessor Lamarck, who admitted at least a general law of progress and development; whereas, according to Darwin, the whole development is due to the gradual summation of innumerable minute and accidental operations."[66]

KARL VOGT

In his preface to his work on the *Descent of Man*, Mr. Darwin quotes [Karl Vogt] as a high authority. We see him elsewhere referred to as one of the first physiologists of Germany. Vogt devotes

---

65. *Lay Sermons*, p. 331.
66. *Sechs Vorlesungen über die Darwinische Theorie* [*Six Lectures on the Darwinian Theory*]. By Ludwig Büchner. Second edition. Leipzig, 1868, vol. 1, p. 125.

the concluding lecture of the second volume of his work on Man to the consideration of Darwinism. He expresses his opinion of it, after high commendation, in the following terms. He says that it cannot be doubted that Darwin's

> theory turns the Creator—and his occasional intervention in the revolutions of the earth and in the production of species—without any hesitation out of doors, inasmuch as it does not leave the smallest room for the agency of such a Being. The first living germ being granted, out of it the creation develops itself progressively by natural selection, through all the geological periods of our planets, by the simple law of descent—no new species arises by creation and none perishes by divine annihilation—the natural course of things, the process of evolution of all organisms and of the earth itself, is of itself sufficient for the production of all we see. Thus Man is not a special creation, produced in a different way, and distinct from other animals, endowed with an individual soul and animated by the breath of God; on the contrary, Man is only the highest product of the progressive evolution of animal life springing from the group of apes next below him.[67]

After this no one can be surprised to hear him say that "the pulpits of the orthodox, the confessionals of the priests, the platforms of the interior missions, the presidential chairs of the consistories, resound with protestations against the assaults made by Materialism and Darwinism against the very foundations of society" (p. 286). This he calls *das Wehgeschrei der Moralisten* (the wail of the Moralists). The designation *Moralists* is a felicitous one as applied to the opponents of Vogt and his associates. It distinguishes them as men who have not lost their moral sense, who refuse to limit their faith to what can be proved by the five senses, who bow to the authority of the law written by the finger of God on the hearts of men, which neither sophistry nor wickedness can effectually erase. All Vogt thinks it necessary to reply to these Moralists is, "*Lässt sie bellen, bis sie ausgebellt haben* (Let them bark till they are tired). *Ende.*"

ERNST HAECKEL

Dr. Ernst Haeckel, Professor in the University of Jena, is said to

---

67. *Vorlesungen über den Menschen, seine Stellung in der Schöpfung und in der Geschichte der Erde* [*Lectures on Humanity, Its Place in the Creation and in the History of the Earth*]. By Karl Vogt. Giessen, 1863, vol. 2, p. 260.

stand at the head of the living naturalists of Germany. His work *Natural History of Creation* contains a course of lectures delivered to the professors, students, and citizens of Jena. It is, therefore, somewhat popular in its character. The ability of the writer is manifest on every page. The distinctness of his perceptions, precision of language, perspicuity of style, and the strength of his convictions give the impression of a man fully master of his subject who has thought himself through and is perfectly satisfied with the conclusions at which he has arrived. At the same time it is the impression of a man who is developed only on one side, who never looks within, who takes no cognizance of the wonders revealed in consciousness, to whom the intuitions of reason and of the conscience, the sense of dependence on a will higher than our own, the sense of obligation and responsibility are of no account—in short a man to whom the image of God enstamped on the soul of man is invisible. This being the case, he that is least in the kingdom of heaven is greater than he.

Haeckel admits that the title of his book, *Natural Creation,* i.e., creation by natural laws, is a contradiction. He distinguishes, however, between the creation of substance and the creation of form. Of the former he says science knows nothing. To the scientist matter is eternal. If anyone chooses to assume that it was created by an extramundane power, Haeckel says he will not object. But that is a matter of faith, and "where faith begins, science ends." The very reverse of this is true. Science must begin with faith. It cannot take a single step without it. How does Haeckel know that his senses do not deceive him? How does he know that he can trust to the operations of his intellect? How does he know that things are as they appear? How does he know that the universe is not a great phantasmagoria, as so many men have regarded it, and man the mere sport of chimeras? He must believe in the laws of belief impressed on his nature. Knowledge implies a mind that knows, and confidence in the act of knowing implies belief in the laws of mind. "An inductive science of nature," says President Porter, "presupposes a science of induction, and a science of induction presupposes a science of man."[68] Haeckel, however, says faith is the mere product of the poetic imag-

---

68. *The Science of Nature versus the Science of Man.* By Noah Porter, President of Yale College. [New York, 1871, p. 29. Porter (1811–92) attempted during his tenure as Yale's president to fuse the college's heritage of commonsense philosophy and Christian belief with a cautious acceptance of the new sciences.]

ination; science, of the understanding; if its conclusions come into conflict with the creations of the imagination, the latter, of course, must give way.[69] He says there have ever been two conflicting theories of the universe: the one monistic, the other dualistic. The one admits of only one substance, matter; the other of two, matter and mind. He prefers to call the former monism rather than materialism because the latter term often includes the idea of moral materialism, i.e., the doctrine that sensual pleasure is the end of life, a doctrine, he says, much more frequently held by princely churchmen than by men of science. He maintains, however, that

> all knowable nature is one; that the same eternal, immutable *(ehernen,* brazen) laws are active in the life of animals and plants, in the formation of crystals, and the power of steam; in the whole sphere of biology, zoology, and botany. We have, therefore, the right to hold fast the monistic and mechanical view, whether men choose to brand the system as Materialism or not. In this sense, all natural science, with the law of causation at its head, is thoroughly materialistic. (p. 32)

The monistic theory he calls "mechanical or causal" as distinguished from the dualistic theory, which he calls "teleological or vitalistic." According to the latter,

> the vegetable and animal kingdoms are considered as the products of a creative agency, working with a definite design. In looking on an organism, the conviction seems unavoidable that so skilfully constructed a machine, such a complicated working apparatus, as an organism is, could be produced only by an agency analogous to, although far more perfect than the agency of man. This supposes the Creator to be an organism analogous to man, although infinitely more perfect; who contemplates his formative powers, lays the plan of the machine, and then, by the use of appropriate means, produces an effect answering to the preconceived plan. . . . However highly the Creator may be exalted, this view involves the ascription to Him of human attributes, in virtue of which He can form a plan, and construct organisms to correspond with it. That is the view to which Dar-

---

69. *Natürliche Schöpfungsgeschichte* [*Natural History of Creation*]. By Dr. Ernst Haeckel, Professor at the University of Jena. Second edition. Berlin, 1873, pp. 8–9.

win's doctrine is directly opposed, and of which Agassiz[70] is, among naturalists, the most important advocate. The famous work of Agassiz, "Essay on Classification," which is in direct opposition to Darwin's, and appeared about the same time, has carried out logically to the utmost the absurd anthropomorphic doctrine of a Creator. (p. 17)

The monistic theory is called "mechanical and causal" because it supposes that all the phenomena of the universe, organic and inorganic, vegetable and animal, vital and mental, are due to mechanical or necessarily operating causes (*causae efficientes*); just as the dualistic theory is called "teleological or vitalistic" because it refers natural organisms to causes working for the accomplishment of a given end (*causae finales*) (p. 67).

The grand difficulty in the way of the mechanical or monistic theory was the occurrence of innumerable organisms, apparently at least, indicative of design. To get over this difficulty, Haeckel says, some who could not believe in a creative and controlling mind adopted the idea of a metaphysical ghost called vitality. The grand service rendered by Darwin to science is that his theory enables us to account for the appearances of design in nature without assuming final causes or a mind working for a foreseen and intended end. "All that had appeared before Darwin," he says, "failed to secure success, and to meet with general acceptance of the doctrine of the mechanical production of vegetable and animal organisms. This was accomplished by Darwin's theory" (p. 20).

The precise difficulty which Mr. Darwin's doctrine has, according to Haeckel, enabled men of science to surmount is thus clearly stated on p. 633. It is "that organs for a definite end should be produced by undesigning or mechanical causes." This difficulty is overcome by the doctrine of evolution:

> Through the theory of descent, we are for the first time able to establish the monistic doctrine of the unity of nature, that a mechanic-causal explanation of the most complicated organisms, *e.g.* the for-

---

70. [Jean Louis Agassiz (1807–73), Swiss-born immigrant to the United States, was Darwin's leading opponent in America. By insisting that species are permanent representations of a divine idea, Agassiz became a favorite of conservatives like Hodge who shared his disagreement with Darwin. Ironically for the orthodox party, however, Agassiz also believed that humanity itself was made up of multiple species, each the product of divine creation.]

mation and constitution of the organs of sense, have no more difficulty for the common understanding, than the mechanical explanation of any physical process, as, for example, earthquakes, the direction of the winds, or the currents of the sea. We thus arrive at the conviction of the last importance, that all natural bodies with which we are acquainted are equally endowed with life (*gleichmässig belebt sind*); that the distinction between living and dead matter does not exist. When a stone is thrown into the air and falls by certain laws to the ground, or when a solution of salt forms a crystal, the result is neither more nor less a mechanical manifestation of life, than the flowering of a plant, the generation or sensibility of animals, or the feelings or the mental activity of man. In thus establishing the monistic theory of nature lies the highest and most comprehensive merit of the doctrine of descent, as reformed by Darwin. . . .

As to the much vaunted design in nature, it is a reality only for those whose views of animal and vegetable life are to the last degree superficial. Any one who has gone deeper into the organization and vital activity of animals and plants, who has made himself familiar with the action and reaction of vital phenomena, and the so-called economy of nature, comes of necessity to the conclusion, that design does not exist, any more than the vaunted goodness of the Creator (*die vielgerühmte Allgüte des Schöpfers*). (pp. 21, 17)

Professor Huxley in his review of this work of Haeckel, already quoted, says: "I do not like to conclude without reminding the reader of my entire concurrence with the general tenor and spirit of the work, and of my high estimate of its value." If you take out of Haeckel's book its doctrine of Monism, which he himself says means Materialism, it has no "tenor or spirit" in it. It is not, however, for us to say how far Professor Huxley intended his indorsement to go.

Haeckel says that Darwin's theory of evolution leads inevitably to Atheism and Materialism. In this we think he is correct. But we have nothing to do with Haeckel's logic or with our own. We make no charge against Mr. Darwin. We cite Haeckel merely as a witness to the fact that Darwinism involves the denial of final causes, that it excludes all intelligent design in the production of the organs of plants and animals, and even in the production of the soul and body of man. This first of German naturalists would occupy a strange position in the sight of all Europe if, after lauding a book to the skies because it teaches a certain doctrine, it should turn out that the book taught no such doctrine at all.

Testimony of the Opponents of Darwinism

THE DUKE OF ARGYLL

When cultivated men undertake to refute a certain system, it is to be presumed that they give themselves the trouble to ascertain what that system is. As the advocates of Mr. Darwin's theory defend and applaud it because it excludes design, and as its opponents make that the main ground of their objection to it, there can be no reasonable doubt as to its real character. The question is, How are the contrivances in nature to be accounted for? One answer is, They are due to the purpose of God. Mr. Darwin says, They are due to the gradual and undesigned accumulation of slight variations. The Duke's first objection to that doctrine is that the evidence of design in the organs of plants and animals is so clear that Mr. Darwin himself cannot avoid using teleological language. "He exhausts," he says,

> every form of words and of illustration by which intention or mental purpose can be described. "Contrivance," "beautiful contrivance," "curious contrivance," are expressions which occur over and over again. Here is one sentence describing a particular species (of orchids): "The labellum is developed *in order* to attract the Lepidoptera [butterflies]; and we shall soon see reason for supposing that the nectar is purposely so lodged, that it can be sucked only slowly *in order* to give time for the curious chemical quality of the matter setting hard and dry."[71]

We have already seen that Mr. Darwin's answer to this objection is that it is hard to keep from personifying nature, and that these expressions as used by him mean no more than chemists mean when they speak of affinities and one element preferring another.

A second objection is that a variation would not be useful to the individual in which it happens to occur unless other variations should occur at the right time and in the right order, and that the concurrence of so many accidents as are required to account for the infinite diversity of forms in plants and animals is altogether inconceivable.

A third objection is that the variations often have no reference to the organism of the animal itself but to other organisms. "Take one

---

71. [The Duke of Argyll.] *Reign of Law*. London, 1867, p. 40.

instance," he says, "out of millions. The poison of a deadly snake—let us for a moment consider what that is. It is a secretion of definite chemical properties with reference not only—not even mainly—to the organism of the animal in which it is developed, but specially to another animal which it is intended to destroy." "How," he asks,

> will the law of growth adjust a poison in one animal with such subtle knowledge of the organization of the other, that the deadly virus shall in a few minutes curdle the blood, benumb the nerves, and rush in upon the citadel of life? There is but one explanation: a Mind having minute and perfect knowledge of the structure of both has designed the one to be capable of inflicting death upon the other. This mental purpose and resolve is the one thing which our intelligence perceives with direct and intuitive recognition. The method of creation by which this purpose has been carried into effect is utterly unknown.[72]

A fourth objection has reference to beauty. According to Mr. Darwin, flowers are not intentionally made beautiful, but those which happen to be beautiful attract insects, and by their agency are fertilized and survive. Male birds are not intentionally arrayed in bright colors, but those which happen to be so arrayed are attractive, and thus become the progenitors of their race. Against this explanation the Duke earnestly protests. He refers to the gorgeously adorned class of humming-birds, of which naturalists enumerate no less than four hundred and thirty different species, distinguished one from the other, in general, only by their plumage. "Now," he asks,

> what explanation does the law of natural selection give—I will not say of the origin, but even of the continuance of such specific varieties as these? None whatever. A crest of topaz is no better in the struggle of existence than a crest of sapphire. A frill ending in spangles of the emerald is no better in the battle of life than a frill ending in spangles of the ruby. A tail is not affected for the purposes of flight, whether its marginal, or its central feathers are decorated with white. It is impossible to bring such varieties into any physical law known to us. It has relation however to a Purpose, which stands in close analogy with our knowledge of purpose in the works of men. Mere beauty and mere variety, for their own sake, are objects which we ourselves seek, when we can make the forces of nature subordinate to the attainment

---

72. *Reign of Law*. London, 1867, p. 37.

of them. There seems to be no conceivable reason why we should doubt or question that these are ends and aims also in the forms given to living organisms, when the facts correspond with this view and with no other.[73]

It will be observed that all these objections have reference to the denial of teleology on the part of Mr. Darwin. If his theory admitted that the organisms in nature were due to a divine purpose, the objections would be void of all meaning.

There is a fifth objection. According to Darwin's theory, organs are formed by the slow accumulation of unintended variations which happen to be favorable to the subject of them in the struggle for life. But in many cases these organs, instead of being favorable, are injurious or cumbersome until fully developed. Take the wing of a bird, for example. In its rudimental state, it is useful neither for swimming, walking, nor flying. Now, as Darwin says it took millions of years to bring the eye to perfection, how long did it take to render a rudimental wing useful? It is no sufficient answer to say that these rudimental organs might have been suited to the condition in which the animal existed during the formative process. This is perfectly arbitrary. It has no basis of fact. There are but three kinds of locomotion that we know of: in the water, on the ground, and through the air; for all these purposes a half-formed wing would be an impediment.

The Duke devotes almost a whole chapter of his interesting book to the consideration of "contrivance in the machinery for flight." The conditions to secure regulated movement through the atmosphere are so numerous, so complicated, and so conflicting, that the problem never has been solved by human ingenuity. In the structure of the bird it is solved to perfection. As we are not writing a teleological argument, but only producing evidence that Darwinism excludes teleology, we cannot follow the details which prove that the wing of the gannet or swift is almost as wonderful and beautiful a specimen of contrivance as the eye of the eagle.

LOUIS AGASSIZ

Every one knows that the illustrious Agassiz, over whose recent grave the world stands weeping, was from the beginning a pro-

---

73. *Reign of Law*, pp. 247–48.

nounced and earnest opponent of Mr. Darwin's theory. He wrote as a naturalist, and therefore his objections are principally directed against the theory of evolution, which he regarded as not only destitute of any scientific basis, but as subversive of the best established facts in zoology. Nevertheless it is evident that his zeal was greatly intensified by his apprehension that a theory which obliterates all evidence of the being of God from the works of nature endangers faith in that great doctrine itself. The Rev. Dr. Peabody,[74] in the discourse delivered on the occasion of Professor Agassiz's funeral, said:

> I cannot close this hasty and inadequate, yet fervent and hearty tribute, without recalling to your memory the reverent spirit in which he pursued his scientific labors. Nearly forty years ago, in his first great work on fossil fishes, in developing principles of classification, he wrote in quotations, "An invisible thread in all ages runs through this immense diversity, exhibiting as a general result that there is a continual progress in development ending in man, the four classes of vertebrates presenting the intermediate steps, and the invertebrates the constant accessory accompaniment. Have we not here the manifestation of a mind as powerful as prolific? an act of intelligence as sublime as provident? the marks of goodness as infinite as wise? the most palpable demonstration of the existence of a personal God, author of all this; ruler of the universe, and the dispenser of all good? This at least is what I read in the works of creation." And it was what he ever read, and with profound awe and adoration. To this exalted faith he was inflexibly loyal. The laws of nature were to him the eternal Word of God.
>
> His repugnance to Darwinism grew in great part from his apprehension of its atheistical tendency—an apprehension which I confess I cannot share; for I forget not that these theories, now in the ascendent, are maintained by not a few devout Christian men, and while they appear to me unproved and incapable of demonstration, I could admit them without parting with one iota of my faith in God and Christ. Yet I cannot but sympathize most strongly with him in the spirit in which he resisted what seemed to him lese-majesty against the sovereign of the universe. Nor was his a theoretical faith. His whole life, in its broad philanthropy, in its pervading spirit of service, in its fidelity to arduous trusts and duties, and in its simplicity and

---

74. [Andrew Peabody (1811–93) was a conservative Unitarian minister and twice the acting president of Harvard.]

truthfulness, bespoke one who was consciously fulfilling a mission from God to his fellow-men.

The words *evolution* and *Darwinism* are so often in this country, but not in Europe, used interchangeably that it is conceivable that Dr. Peabody could retain his faith in God, and yet admit the doctrine of evolution. But it is not conceivable that any man should adopt the main element of Mr. Darwin's theory, viz., the denial of all final causes and the assertion that since the first creation of matter and life God has left the universe to the control of unintelligent physical causes, so that all the phenomena of the plants and animals, all that is in man, and all that has ever happened on the earth, is due to physical force, and yet retain his faith in Christ. On that theory, there have been no supernatural revelation, no miracles; Christ is not risen, and we are yet in our sins. It is not thus that this matter is regarded abroad. The Christians of Germany say that the only alternative these theories leave us is Heathenism or Christianity: *Heidenthum oder Christenthum, die Frage der Zeit* [Heathenism or Christianity, the question of the day].

PAUL JANET

Paul Janet, a professor of philosophy, is the author of a book on the Materialism of Büchner.[75] The greater part of the last chapter of his work is devoted to Darwinism. He says, "Dr. Büchner invoked (Darwin's book) as a striking confirmation of his doctrine" (p. 154). What Büchner's doctrine is has been shown on a previous page. The points of coincidence between Darwin's system and his are that both regard mind as a mere function of living matter, and both refer all the organs and organisms of living things to the unconscious, unintelligent operation of physical causes. Büchner's way of accounting for complicated organs was "the energy of the elements and forces of matter, which in their fated and accidental occurrence must have produced innumerable forms, which must needs limit each other mutually, and correspond, apparently, the one with the other, as if they were made for that purpose. Out of all those forms,

---

75. *The Materialism of the Present Day: A Critique of Dr. Büchner's System.* By Paul Janet, Member of the Institute of France, Professor of Philosophy at the Paris Faculté des Lettres. Translated from the French, by Gustave Masson, B.A. London and Paris, 1867. [Janet (1823–99) held that spirit could exist distinct from bodies and was, by comparison with materiality, the only real existence.]

they only have survived which were adapted, in some manner, to the conditions of the medium in which they were placed" (p. 30). This is very clumsy. No wonder Büchner preferred Darwin's method. The two systems are, indeed, exactly the same, but Mr. Darwin has a much more winning way of presenting it.

Professor Janet does not seem to have much objection to the doctrine of evolution in itself; it is the denial of teleology that he regards as the fatal element of Mr. Darwin's theory. "According to us," he says, "the true stumbling-block of Mr. Darwin's theory, the perilous and slippery point, is the passage from artificial to natural selection; it is when he wants to establish that a blind and designless nature has been able to obtain, by the occurrence of circumstances, the same results which man obtains by thoughtful and well calculated industry" (p. 174).

Towards the end of his volume he says:

> We shall conclude by a general observation. Notwithstanding the numerous objections we have raised against Mr. Darwin's theory, we do not declare ourselves hostile to a system of which zoologists are the only competent judges. We are neither for nor against the transmutation of species, neither for nor against the principle of natural selection. The only positive conclusion of our debate is this: no principle hitherto known, neither the action of media, nor habit, nor natural selection, can account for organic adaptations without the intervention of the principle of finality. Natural selection, unguided, submitted to the laws of a pure mechanism, and exclusively determined by accidents, seems to me, under another name, the chance proclaimed by Epicurus, equally barren, equally incomprehensible; on the other hand, natural selection guided beforehand by a provident will, directed towards a precise end by intentional laws, might be the means which nature has selected to pass from one stage of being to another, from one form to another, to bring to perfection life throughout the universe, and to rise by a continuous process from the monad to man. Now, I ask Mr. Darwin himself, what interest has he in maintaining that natural selection is not guided—not directed? What interest has he in substituting accidental causes for every final cause? I cannot see. Let him admit that in natural, as well as in artificial selection, there may be a choice and direction; his principle immediately becomes much more fruitful than it was before. His hypothesis, then, whilst having the advantage of exempting science from the necessity of introducing the personal and miraculous intervention of God in the creation of each species, yet would be free from

the banishing out of the universe an all-provident thought, and of submitting everything to blind and brute chance. (pp. 198–99)

Professor Janet asks far too much of Mr. Darwin. To ask him to give up his denial of final causes is like asking the Romanists to give up the Pope. That principle is the life and soul of his system.

M. FLOURENS

M. Flourens,[76] recently dead, was one of the earliest and most pronounced opponents of Darwinism. He published in 1864 his *Examen du Livre de M. Darwin sur l'Origine des Espèces* [*Examination of Darwin's Book on the Origin of Species*]. His positions as Member of the Académie Française, and Perpetual Secretary of the Académie des Sciences, or Institut de France, vouch for his high rank among the French naturalists. His connection with the Jardin des Plantes [Botanical Garden] gave him enlarged opportunities for biological experiments. The result of his own experience, as well as the experience of other observers, was, as he expresses it, his solemn conviction that species are fixed and not transmutable. No ingenuity of device could render hybrids fertile. "They never establish an intermediate species." It is, therefore, to the doctrine of evolution his attention is principally directed.

Nevertheless, he is no less struck by Darwin's way of excluding all intelligence and design in his manner of speaking of nature. On this point he quotes the language of Cuvier,[77] who says:

> Nature has been personified. Living beings have been called the works of nature. The general bearing of these creatures to each other has become the laws of nature. It is thus while considering Nature as a being endowed with intelligence and will, but in its power limited and secondary, that it may be said that she watches incessantly over the maintenance of her work; that she does nothing in vain, and always acts by the most simple means. . . . It is easy to see how puerile are those who give nature a species of individual existence distinct from the Creator, and from the law which He has impressed upon the movements and peculiarities of the forms given by Him to living

---

76. [Marie Jean Pierre Flourens (1794–1867) was a French physiologist and protégé of Baron Cuvier.]
77. [Baron Georges Cuvier (1769–1832), renowned French paleontologist, held that rapid global upheavals account for the variability of species.]

things, and which He makes to act upon their bodies with a peculiar force and reason.

Older writers, says Flourens, in speaking of Nature, "gave to her inclinations, intentions, and views, and horrors (of a vacuum), and sports," etc. He says that one of the principal objects of his book is to show how Mr. Darwin "has deluded himself, and perhaps others, by a constant abuse of figurative language. . . . He plays with Nature as he pleases, and makes her do whatsoever he wishes." When we remember that Mr. Darwin defines Nature to be the aggregate of physical forces, we see how, in attributing everything to Nature, he effectually excludes the supernatural.

In his volume of *Lay Sermons, Reviews,* etc., Professor Huxley has a very severe critique on M. Flourens's book. He says little, however, in reference to teleology, except in one paragraph in which we read: "M. Flourens cannot imagine an unconscious selection; it is for him a contradiction in terms." Huxley's answer is, "The winds and waves of the Bay of Biscay have not much consciousness, and yet they have with great care 'selected,' from an infinity of masses of silex, all grains of sand below a certain size and have heaped them by themselves over a great area. . . . A frosty night selects the hardy plants in a plantation from among the tender ones as effectually as if the intelligence of the gardener had been operative in cutting the weaker ones down."[78] If this means anything, it means that as the winds and waves of the Bay of Biscay can make heaps of sand, so similar unconscious agencies can, if you only give them time enough, make an elephant or a man; for this is what Mr. Darwin says natural selection has done.

REV. WALTER MITCHELL, M.A.,
VICE-PRESIDENT OF THE VICTORIA INSTITUTE

The Victoria Institute, or Philosophical Society of Great Britain,[79] under the presidency of the Earl of Shaftesbury,[80] includes among

---

78. *Lay Sermons,* p. 347.
79. [The Victoria Institute, or Philosophical Society of Great Britain, was founded in 1865 by evangelicals who wished to demonstrate the compatibility between Scripture and science. Rev. Walter Mitchell, vice-president of the institute, wrote in defense of biblical inspiration and on the use of reason to correlate science and religion.]
80. [Anthony Ashley Cooper, seventh Earl of Shaftesbury (1801–85), was famous for his efforts at improving the lot of the poor. He was an ardent evangelical who patronized many Christian organizations like the Victoria Institute.]

its members many of the dignitaries of the Church of England, and a large number of distinguished men of different professions and denominations. Its principal object is "to investigate fully and impartially the most important questions of philosophy and science, but more especially those that bear on the great truths revealed in Holy Scripture, with the view of defending these truths against the opposition of Science, falsely so called." The Institute holds bi-monthly meetings at which papers are read on some important topic and then submitted to criticism and discussion. These papers, many of which are very elaborate, are published in the *Transactions of the Institute,* together with a full report of the discussions to which they gave rise. Six volumes, replete with valuable and varied information, have already been published.

Very considerable latitude of opinion is allowed. Hence we find in the *Transactions* papers for and against evolution—for and against Darwinism. It would be easy to quote extracts, pertinent to our subject, more than enough to fill a volume much larger than the present. We must content ourselves with a few citations from the discussion on a paper in favor of the credibility of Darwinism,[81] and another in favor of the doctrine of evolution.[82] In summing up the debates on these two topics, the chairman, Rev. Walter Mitchell, presented with great clearness and force his reasons for regarding Darwinism as incredible and impossible. In his protracted remarks he contrasts the Scriptural doctrine, that of the *Vestiges of Creation,* and that of Darwin on the origin of species. He thus states the doctrine of the Bible on the subject:

> If science be another name for real knowledge; if science be the pursuit of sound wisdom; if science be the pursuit of truth itself; I say that man has no right to reject anything that is true because it savors of God. Well, what is this hypothesis—older than that of Darwin—which

---

81. "The Credibility of Darwinism." By George Warington, Esq., F.C.S., M.V.I. [This paper was presented by Warington to the Victoria Institute and published in the second volume of its *Proceedings* (1867). Warington averred that Darwinism was "a good working hypothesis," but the Secretary of the Institute, James Reddie, reflected the general feelings of the meeting when he said he was not convinced by Warington's arguments. The editors have found no further information on Warington. "M.V.I." here and in the next note means "Member of the Victoria Institute."]

82. "On Certain Analogies between the Methods of Deity in Nature and Revelation." By Rev. G. E. Henslow, M.A., F.L.S., M.V.I.

does, and does alone, account for all the observed facts, or all that which we can read, recorded in the book of Nature? It is, that God created all things very good; that He made every vegetable after its own kind; that He made every animal after its own kind; that He allowed certain laws of variation, but that He has ordained strict, though invisible and invincible barriers, which prevent that variation from running riot, and which include it within strict and well defined limits. This is a hypothesis which will account for all that we have learnt from the works of Nature. It admits an intelligent Being as the Author of all the works of creation, animate as well as inanimate; it leaves no mysteries in the animate world unaccounted for. There is one thing which the animate, as well as the inanimate world declares to man, one thing everywhere plainly recorded, if we will only read it, and that is the impress of design, the design of infinite wisdom. Any theory which comes in with an attempt to ignore design as manifested in God's creation, is a theory, I say, which attempts to dethrone God. This the theory of Darwin does endeavor to do. If asked how our old theory accounts for such uniformity of design in the midst of such perplexing variety as we find in nature, we reply, that this can only be accounted for on one admission, that the whole is the work of one Author, built according, as it were, to one style; that it represents the unity of one mind with the infinite power of adapting all its works in the most perfect manner for the uses for which they were created. Whewell has boldly maintained, and he has never been controverted, that all real advances in the sciences of physiology and comparative anatomy—such as that made by Harvey[83] in discovering the circulation of the blood—have been made by those who not only believed in the existence of design everywhere manifested in the animate world, but were led by that belief to make their discoveries.

When discussing the paper of Mr. Henslow on evolution [Mitchell] says:

> In speaking of this paper I must commend the exceeding reverent tone in which the author has discussed the subject, and I should like to see all such subjects discussed in a similar tone. The view which Mr. Henslow brings forward, however, does not appear to be a very original one. It was the first view ever brought forward on the doctrine of evolution, and I was the first one to point out that the whole doctrine

---

83. [In certain religious circles during the nineteenth century, William Harvey (1578–1657) came to be respected as much for his Christian faith as for his pioneering discoveries concerning the circulation of the blood.]

was one of retrograde character. The whole tone and character of this paper, except that which relates to the attributes and moral government of God,[84] is nothing more or less than the same view of the doctrine of evolution which created such a sensation in this country when that famous book came out, "The Vestiges of Creation." So far as I can understand the arguments of Mr. Darwin, they have simply been an endeavor to eject out of the idea of evolution the personal work of the Deity. His whole endeavor has been to push the Creator farther and farther back out of view. The most laborious part of Darwin's attempt at reasoning—for it is not true reasoning—the most laborious part of his logic and reasoning, is intended to eliminate, as perfectly as any of the atheistical authors have endeavored to do, the idea of design. Now, setting revelation aside, the manner in which the unknown author of the "Vestiges of Creation" treated this subject, satisfactorily showed that the doctrine of evolution was not in itself an atheistical doctrine, nor did it deny the existence of design. So far as I could understand and make out, having carefully read the book at the time it came out and afterwards, and having carefully analyzed and compared it and Mr. Darwin's book with each other, so far as I could understand it, the doctrine of the author of the "Vestiges of Creation" was simply, that God created all things, and that when He created matter He impressed on it certain laws; that matter, being evolved according to those laws, should produce beings and organs mutually adapted to one another and to the world; and that every successive development which should be produced was essentially foreseen, foreknown, and predetermined by the Deity. His idea, for instance, of the evolution of an eye from a more simple organ was that the ultimate eye—man's eye, for instance—was to be a perfect optical instrument, and that its perfection depended on the previous design by the Creator, that at a certain period it should appear in a body quite adapted for its purposes. There is one question—and not the only one, but we must consider it as an important question—whether you can maintain a doctrine of evolution which shall not be atheistical, and which shall admit the great argument of design? That is one thing; but the next thing is, does such a doctrine as that accord either with revelation or with the facts of science? I do not believe that it can be made to agree with what we believe to be the revealed Word of God, and I do not believe that it has in the least degree been proved that the doctrine is consistent with sound science.

---

84. The second part of Mr. Henslow's paper concerns "the methods of the Deity as revealed to us in the Bible." The same is substantially true of his work *The Theory of Evolution.*

As to Mr. Darwin's theory, it is obvious from the passages already quoted that [Mr. Mitchell] considers its characteristic feature is not evolution, nor even natural selection, but the denial of teleology or of intelligent control.[85] Mr. Darwin admits the original creation of one or a few forms of life; and Mr. Mitchell, in his comments on Mr. Warington's defence of his theory, asks,

> Why am I to limit the work of the Creator to the simultaneous or successive creations of ten or twelve commencements of the animate creation? Why, simply for the purpose of evading the evidence of design as manifested in the adaptation of all the organs of every animate creature to its wants, which can only be done by so incredible an hypothesis as that of Mr. Darwin. I say fearlessly, that any hypothesis which requires us to admit that the formation of such complex organs as the eye, the ear, the heart, the brain, with all their marvellous structures and mechanical adaptations to the wants of the creatures possessing them, so perfectly in harmony, too, with the laws of inorganic matter, affords no evidence of design; that such structures could be built up by gradual chance improvements, perpetuated by the law of transmission, and perfected by the destruction of creatures less favorably endowed, is so incredible, that I marvel to find any thinking man capable of adopting it for a single moment.

It is useless to multiply quotations. Darwinism is never brought up either formally or incidentally that its exclusion of design in the formation of living organisms is not urged as the main objection against the whole theory.

Principal Dawson

Dr. Dawson, as we are informed, is regarded as the first palaeontologist, and among the first geologists, in America. In his *Story of Earth and Man*[86] he passes in review the several geological periods

---

85. [Although Hodge will later bring up his own objections to evolution per se and in support of the fixity of species, he keeps his eyes here on the larger question of teleology or design and does not pursue Mitchell's comments about the Bible and evolution as such, even though Hodge largely agreed with Mitchell.]

86. *The Story of Earth and Man.* By J. W. Dawson, LL.D., F.R.S., F.G.S., Principal and Vice-Chancellor of McGill University, Montreal. Author of *Archaia, Acadian Geology*, etc. Second edition. London, 1873, p. 497. [John William Dawson (1820–99) was particularly important for Hodge, who had tried hard to recruit him for the faculty of the College of New Jersey. While preparing *What Is Darwinism?* Hodge had spoken personally with Dawson at least twice. For those connections see pp. 27–28.]

recognized by geologists, describes as far as knowable the distribution of land and water during each period, and the vegetable and animal productions by which they were distinguished. His book from beginning to end is anti-Darwinian. In common with other naturalists, his attention is directed principally to the doctrine of evolution, which he endeavors to prove is utterly untenable. That Mr. Darwin's theory excludes teleology is everywhere assumed as an uncontroverted and uncontrovertible fact. "The evolutionist doctrine," he says,

> is itself one of the strangest phenomena of humanity. It existed, and most naturally, in the oldest philosophy and poetry, in connection with the crudest and most uncritical attempts of the human mind to grasp the system of nature; but that in our day a system destitute of any shadow of proof, and supported merely by vague analogies and figures of speech, and by the arbitrary and artificial coherence of its own parts, should be accepted as philosophy, and should find able adherents to string on its thread of hypotheses our vast and weighty stores of knowledge, is surpassingly strange. . . . In many respects these speculations are important, and worthy the attention of thinking men. They seek to revolutionize the religious belief of the world, and if accepted would destroy most of the existing theology and philosophy. They indicate tendencies among scientific thinkers, which, though probably temporary, must, before they disappear, descend to lower strata, and reproduce themselves in grosser forms, and with most serious effects on the whole structure of society. With one class of minds they constitute a sort of religion, which so far satisfies the craving for truth higher than those which relate to immediate wants and pleasures. With another and perhaps larger class, they are accepted as affording a welcome deliverance from all scruples of conscience and fears of a hereafter. In the domain of science evolutionism has like tendencies. It reduces the position of man, who becomes a descendant of inferior animals, and a mere term in a series whose end is unknown. It removes from the study of nature the ideas of final cause and purpose; and the evolutionist, instead of regarding the world as a work of consummate plan, skill, and adjustment, approaches nature as he would a chaos of fallen rocks, which may present forms of castles, and grotesque profiles of men and animals, but they are all fortuitous and without significance. . . .
>
> Taking, then, this broad view of the subject, two great leading alternatives are presented to us. Either man is an independent product of the will of a Higher Intelligence, acting directly or through the

laws and materials of his own institution and production, or he has been produced by an unconscious evolution from lower things. It is true that many evolutionists, either unwilling to offend, or not perceiving the logical consequences of their own hypothesis, endeavor to steer a middle course, and to maintain that the Creator has proceeded by way of evolution. But the bare, hard logic of Spencer, the greatest English authority on evolution, leaves no place for this compromise, and shows that the theory, carried out to its legitimate consequences, excludes the knowledge of a Creator and the possibility of his work. We have, therefore, to choose between evolution and creation, bearing in mind, however, that there may be a place in nature for evolution, properly limited, as well as for other things, and that the idea of creation by no means excludes law and second causes. . . .

It may be said, that evolution may be held as a scientific doctrine in connection with a modified belief in creation. The work of actual creation may have been limited to a few elementary types, and evolution may have done the rest. Evolutionists may still be theists. We have already seen that the doctrine, as carried out to its logical consequences, excludes creation and theism. It may, however, be shown that even in its more modified form, and when held by men who maintain that they are not atheists, it is practically atheistic, because excluding the idea of plan and design, and resolving all things into the action of unintelligent forces. It is necessary to observe this, because it is the half-way-evolutionism, which professes to have a creator somewhere behind it, that is most popular; though it is, if possible, more unphilosophical than that which professes to set out with absolute and determined nonentity, or from self-existing stardust containing all the possibilities of the universe. (pp. 317–18, 321, 340)

In reference to the objection of evolutionists that the origin of every new species, on the theistic doctrine, supposes "a miracle," an intervention of the divine efficiency without the agency of second causes, Principal Dawson asks, "What is the actual statement of the theory of creation as it may be held by a modern man of science? Simply this: that all things have been produced by the Supreme Creative will, acting either directly, or through the agency of the forces and material of his own production" (p. 340).

He thus sums up his argument against the doctrine of evolution, specially in its application to man:

> Finally, the evolutionist picture wants [i.e., lacks] some of the fairest lineaments of humanity, and cheats us with the semblance of man

without the reality. Shave and paint your ape as you may, clothe him and set him up upon his feet, still he fails greatly of the "human form divine"; and so it is with him morally and spiritually as well. We have seen that he wants the instinct of immortality, the love of God, the mental and spiritual power of exercising dominion over the earth. The very agency by which he is evolved is of itself subversive of all these higher properties; the struggle for existence is essentially selfish, and, therefore, degrading. Even in the lower animals, it is a false assumption that its tendency is to elevate; for animals, when driven to the utmost verge of the struggle for life, become depauperated and degraded. The dog which spends its life in snarling contention with its fellow curs for insufficient food, will not be a noble specimen of its race. God does not so treat his creatures. There is far more truth to nature in the doctrine which represents Him as listening to the young ravens when they cry for food. But as applied to man, the theory of the struggle for existence, and survival of the fittest, though the most popular phase of evolutionism at present, is nothing less than the basest and most horrible of superstitions. It makes man not merely carnal but devilish. It takes his lowest appetites and propensities, and makes them his God and Creator. His higher sentiments and aspirations, his self-denying philanthropy, his enthusiasm for the good and true, all the struggles and sufferings of heroes and martyrs, not to speak of that self-sacrifice which is the foundation of Christianity, are, in the view of the evolutionist, mere loss and waste, failure in the struggle of life. What does he give us in exchange? An endless pedigree of bestial ancestors, without one gleam of high and holy tradition to enliven the procession; and for the future, the prospect that the poor mass of protoplasm, which constitutes the sum of our being, and which is the sole gain of an indefinite struggle in the past, must soon be resolved again into inferior animals or dead matter. That men of thought and culture should advocate such a philosophy, argues either a strange mental hallucination, or that the higher spiritual nature has been wholly quenched within them. It is one of the saddest of many sad spectacles which our age presents. (p. 395)

Relation of Darwinism to Religion

Causes of the Alienation between Science and Religion

The consideration of [the relation between Darwinism and religion] would lead into the wide field of the relation between science and religion. Into that field we lack competency and time to enter; a few remarks, however, on the subject may not be out of place.

Those remarks we would fain make in a humble way irenical. There is need of an Irenicum,[87] for the fact is painfully notorious that there is an antagonism between scientific men as a class and religious men as a class. Of course this opposition is neither felt nor expressed by all on either side. Nevertheless, whatever may be the cause of this antagonism, or whoever are to be blamed for it, there can be no doubt that it exists and that it is an evil.

The first cause of the alienation in question is that the two parties, so to speak, adopt different rules of evidence, and thus can hardly avoid arriving at different conclusions. To understand this we must determine what is meant by science and by scientific evidence. Science, according to its etymology, is simply knowledge. But usage has limited its meaning, in the first place, not to the knowledge of facts or phenomena merely, but to their causes and relations. It was said of old, "ὅτι *scientiae fundamentum*, διότι *fastigium* [Because knowledge is the foundation, it is therefore the pinnacle too]." No amount of materials would constitute a building. They must be duly arranged so as to make a symmetrical whole. No amount of disconnected data can constitute a science. Those data must be systematized in their relation to each other and to other things.

In the second place, the word is becoming more and more restricted to the knowledge of a particular class of facts and of their relations, namely, the facts of nature or of the external world. This usage is not universal, nor is it fixed. In Germany, especially, the word *Wissenschaft* is used of all kinds of ordered knowledge, whether transcendental or empirical. So we are accustomed to speak of mental, moral, social, as well as of natural science.[88] Nevertheless, the more restricted use of the word is very common and very influential.

It is important that this fact should be recognized. In common usage, a scientific man is distinguished specially from a metaphysician. The one investigates the phenomena of matter, the other studies the phenomena of mind, according to the old distinction between physics and metaphysics. Science, therefore, is the ordered

---

87. [That is, a means of establishing peace. At this point Hodge carries on with a concern that had been a major part of his entire intellectual life, namely, the unity of sciences, including theology.]

88. [These are the terms used by the College of New Jersey in describing its curriculum when Hodge was an undergraduate. Hodge's defense of the dignity of other forms of knowledge alongside information from sense data continues a major theme of his theological career.]

knowledge of the phenomena which we recognize through the senses. A scientific fact is a fact perceived by the senses. Scientific evidence is evidence addressed to the senses. At one of the meetings of the Victoria Institute, a visitor avowed his disbelief in the existence of God. When asked what kind of evidence would satisfy him, he answered, Just such evidence as I have of the existence of this tumbler which I now hold in my hand. The Rev. Mr. Henslow says, "By science is meant the investigation of facts and phenomena recognizable by the senses, and of the causes which have brought them into existence."[89]

This is the main root of the trouble. If science be the knowledge of the facts perceived by the senses, and scientific evidence, evidence addressed to the senses, then the senses are the only sources of knowledge. Any conviction resting on any other ground than the testimony of the senses must be faith. Darwin admits that the contrivances in nature may be accounted for by assuming that they are due to design on the part of God. But, he says, that would not be science. Haeckel says that to science matter is eternal. If any man chooses to say it was created, well and good; but that is a matter of faith, and faith is imagination. Ulrici quotes a distinguished German physiologist who believes in vital as distinguished from physical forces, but he holds to spontaneous generation, not, as he admits, because it has been proved, but because the admission of any higher power than nature is unscientific.[90]

It is inevitable that minds addicted to scientific investigation should receive a strong bias to undervalue any other kind of evidence except that of the senses, *i.e.*, scientific evidence. We have seen that those who give themselves up to this tendency come to deny God, to deny mind, to deny even self. It is true that the great majority of men, scientific as well as others, are so much under the control of the laws of their nature that they cannot go to this extreme. The tendency, however, of a mind addicted to the consideration of one kind of evidence to become more or less insensible to other kinds of proof is undeniable. Thus even Agassiz, as a zoologist and simply on zoological grounds, assumed that there were sev-

---

89. "Science and Scripture Not Antagonistic, Because Distinct in Their Spheres of Thought." A Lecture, by Rev. George Henslow, M.A., F.L.S., F.G.S. London, 1873, p. 1.

90. *Gott und Natur*, p. 200.

eral zones between the Ganges and the Atlantic Ocean, each having its own flora and fauna, and inhabited by races of men the same in kind, but of different origins. When told by the comparative philologists that this was impossible because the languages spoken through that wide region demonstrated that its inhabitants must have had a common descent, he could only answer that as ducks quack everywhere, he could not see why men should not everywhere speak the same language.

A still more striking illustration is furnished by Dr. Lionel Beale,[91] the distinguished English physiologist. He has written a book of three hundred and eighty-eight pages for the express purpose of proving that the phenomena of life, instinct, and intellect cannot be referred to any known natural forces. He avows his belief that in nature "mind governs matter," and "in the existence of a never-changing, all-seeing, power-directing and matter-guiding Omnipotence." He avows his faith in miracles and "those miracles on which Christianity is founded." Nevertheless, his faith in all these points is provisional. He says that a truly scientific man, "if the maintenance, continuity, and nature of life on our planet should at some future time be fully explained without supposing the existence of any such supernatural omnipotent influence, would be bound to receive the new explanation, and might abandon the old conviction."[92] That is, all evidence of the truths of religion not founded on nature and perceived by the senses amounts to nothing.

Now as religion does not rest on the testimony of the senses, that is, on scientific evidence, the tendency of scientific men is to ignore its claims. We speak only of tendency. We rejoice to know or believe that in hundreds or thousands of scientific men, this tendency is counteracted by their consciousness of manhood—the conviction that the body is not the man—by the intuitions of the reason and the conscience, and by the grace of God. No class of men stands deservedly higher in public estimation than men of science who, while remaining faithful to their higher nature, have enlarged our knowledge of the wonderful works of God.

91. [Lionel Smith Beale (1828–1906), an English physiologist noted for his work with the microscope, vigorously opposed Huxley and mechanistic interpretations of life by contending for what he called a "vital force" as the foundation of existence.]

92. *Protoplasm; or, Matter and Life.* By Lionel S. Beale, M.B., F.R.S. Third edition. London & Philadelphia, 1874, p. 345, and the whole chapter on design.

A second cause of the alienation between science and religion is the failure to make the due distinction between facts and the explanation of those facts, or the theories deduced from them.[93] No sound-minded man disputes any scientific fact. Religious men believe with Agassiz that facts are sacred. They are revelations from God. Christians sacrifice to them, when duly authenticated, their most cherished convictions. That the earth moves no religious man doubts. When Galileo made that great discovery, the Church was right in not yielding at once to the evidence of an experiment which it did not understand. But when the fact was clearly established, no man set up his interpretation of the Bible in opposition to it. Religious men admit all the facts connected with our solar system, all the facts of geology, and of comparative anatomy, and of biology. Ought not this to satisfy scientific men? Must we also admit their explanations and inferences? If we admit that the human embryo passes through various phases, must we admit that man was once a fish, then a bird, then a dog, then an ape, and finally what he now is? If we admit the similarity of structure in all vertebrates, must we admit the evolution of one from another, and all from a primordial germ? It is to be remembered that the facts are from God, the explanation from men; and the two are often as far apart as Heaven and its antipode.

These human explanations are not only without authority, but they are very mutable. They change not only from generation to generation, but almost as often as the phases of the moon. It is a fact that the planets move. Once it was said that they were moved by spirits, then by vortexes, now by self-evolved forces. It is hard that we should be called upon to change our faith with every new moon.

The same man sometimes propounds theories almost as rapidly as the changes of the kaleidoscope. The amiable Sir Charles Lyell,[94]

---

93. [This discussion of the difference between "facts" and "explanations" repeats a consistent theme in Hodge's work. See, for example, *Systematic Theology*, 3 vols. (New York: Scribner, 1872–73), 1:13: "The investigator sees, or ascertains by observation, what are the laws which determine material phenomena; he does not invent those laws. His speculations on matters of science unless sustained by facts, are worthless."]

94. [Charles Lyell (1797–1875), most influential geologist of his age, believed that geological change took place at a uniform rate in time. In the last edition of his famous *Principles of Geology*, under the personal as well as professional influence of Darwin, Lyell accepted a modified version of the transmutation of species.]

England's most distinguished geologist, has published ten editions of his *Principles of Geology,* which so differ as to make it hard to believe that it is the work of the same mind. "In the eighth edition of his work," says Dr. Bree,

> Sir Charles Lyell, the Nestor of geologists, to whom the present generation is more indebted than to any other for all that is known of geology in its advanced stage, teaches that species have a real existence in nature, and that each was endowed at the time of its creation with the attributes and organization by which it is now distinguished. ... In all the editions up to the tenth, he looked upon geological facts and geological phenomena as proving the fixity of species and their special creation in time. In the tenth edition, just published, he announces his change of opinion on this subject and his conversion to the doctrine of development by law.[95]

The change on the part of this eminent geologist, it is to be observed, is a mere change of opinion. There was no change of the facts of geology between the publication of the eighth and of the tenth edition of his work, neither was there any change in his knowledge of those facts. All the facts relied upon by evolutionists have long been familiar to scientific men. The whole change is a subjective one. One year the veteran geologist thinks the facts teach one thing, another year he thinks they teach another. It is now the fact, and it is feared it will continue to be a fact, that scientific men give the name of science to their explanations as well as to the facts. Nay, they are often, and naturally, more zealous for their explanations than they are for the facts. The facts are God's, the explanations are their own.

The third cause of the alienation between religion and science is the bearing of scientific men towards the men of culture who do not belong to their own class. When we, in such connections, speak of scientific men, we do not mean men of science as such, but those only who avow or manifest their hostility to religion. There is an assumption of superiority, and often a manifestation of contempt. Those who call their logic or their conjectures into question are

---

95. *Fallacies in the Hypothesis of Mr. Darwin,* by C. R. Bree, M.D., F.Z.S. London, 1872, p. 290. [Charles Robert Bree (1811–86), a physician, was one of Darwin's most philosophically minded critics; he held that proofs for the fixity of species were as numerous as, and more persuasive than, those for the fluidity of species.]

stigmatized as narrow-minded, bigots, old women, Bible worshippers, etc.

Professor Huxley's advice to metaphysicians and theologians is to let science alone. This is his Irenicum. But do he and his associates let metaphysics and religion alone? They tell the metaphysician that his vocation is gone, there is no such thing as mind, and of course no mental laws to be established. Metaphysics are merged into physics. Professor Huxley tells the religious world that there is overwhelming and crushing evidence (scientific evidence, of course) that no event has ever occurred on this earth which was not the effect of natural causes. Hence there have been no miracles, and Christ is not risen.[96] He says that the doctrine that belief in a personal God is necessary to any religion worthy of the name is a mere matter of opinion. Tyndall, Carpenter, and Henry Thompson[97] teach that prayer is a superstitious absurdity; Herbert Spencer, whom they call their "great philosopher," *i.e.*, the man who does their thinking, labors to prove that there cannot be a personal God or human soul or self, that moral laws are mere "generalizations of utility," or, as Karl Vogt says, that self-respect, and not the will of God, is the ground and rule of moral obligation.

If any protest be made against such doctrines, we are told that scientific truth cannot be put down by denunciation (or as Vogt says, by barking). So the Pharisees, when our blessed Lord called them hypocrites and a generation of vipers, and said, "Ye compass sea and land to make one proselyte, and when he is made, ye make him twofold more the child of hell than yourselves" [Matt. 23:15], doubtless thought that that was a poor way to refute their theory that holiness and salvation were to be secured by church-membership and church-rites. Nevertheless, as those words were the words of Christ, they were a thunderbolt which reverberates through all time and space, and still makes Pharisees of every name and nation tremble.

---

96. When Professor Huxley says that he does not deny the possibility of miracles [see p. 105, n. 60], he must use the word *miracle* in a sense peculiar to himself.

97. [William B. Carpenter (1813–85), physician, physiologist, and naturalist, lectured in medicine at the University of London. From a Unitarian theological position he came to accept natural selection as an explanation for what seemed to be a divinely ordered creation. Sir Henry Thompson (1820–1904) was a notable surgeon who promoted cremation. He was also renowned for proposing a "prayer experiment": to test the efficacy of divine influences in the world, selected patients in a hospital would become the object of prayer for healing.]

Huxley's Irenicum will not do. Men who are assiduously poisoning the fountains of religion, morality, and social order cannot be let alone. Haeckel's Irenicum amounts to much the same as that of Professor Huxley. The right to speak on these vital subjects he forbids to all who are not thoroughly versed in biology, and who are not entirely emancipated from the trammels of their long cherished traditional beliefs.[98] This, as the whole context shows, means that a man in order to be entitled to be heard on the evolution theory must be willing to renounce his faith not only in the Bible, but in God, in the soul, in a future life, and become a monistic materialist.[99]

98. *Jenaer Literaturzeitung* [*Jena Literary Magazine*], January 3, 1874. In this number there is a notice by Doctor Haeckel of two books—*Descendenzlehre und Darwinismus* [*Inheritance Theory and Darwinism*] by Oskar Schmidt, Leipzig, 1873; and *Die Fortschritte des Darwinismus* [*The Progress of Darwinism*] by J. W. Spengel, Cologne and Leipzig, 1874—in which he says: "Erstens, um in Sachen der Descendenz-Theorie mitreden zu können, ein gewisser Grad von tieferer biologischer (sowohl morphologischer als physiologischer) Bildung unentbehrlich ist, den die meisten von jenen Auctoren (the opposers of the theory) nicht besitzen. Zweitens ist für ein klares und zutreffendes Urtheil in diesen Sachen eine rücksichtslose Hingabe an vernunftgemässe Erkenntnis und eine dadurch bedingte Resignation auf uralte, liebgewordene und tief vererbte Vorurtheile erforderlich, zu welcher sich die wenigsten entschliessen können. [First, in order to participate in the discussion of inheritance theory, it is indispensable to have a certain level of solid training in biology (not to speak of morphology and physiology), which most of the authors (the opposers of the theory) do not have. Second, for a clear and compelling view of this matter it is necessary to abandon oneself ruthlessly to judgments of reason and to hold in abeyance ancient, beloved, and deeply ingrained judgments. Only a very few are able to commit themselves to such discipline.]"

99. In his *Natürliche Schöpfungsgeschichte* [*Natural History of Creation*], Haeckel is still more exclusive. When he comes to answer the objections to the evolution or, as he commonly calls it, the descendence theory, he dismisses the objections derived from religion as unworthy of notice with the remark that *alle Glaube ist Aberglaube;* all faith is superstition. The objections from *a priori* or intuitive truths are disposed of in an equally summary manner by denying that there are any such truths, and asserting that all our knowledge is from the senses. The objection that so many distinguished naturalists reject the theory he considers more at length. First, many have grown old in another way of thinking and cannot be expected to change. Second, many are collectors of facts without studying their relations, or are destitute of the genius for generalization. No amount of material makes a building. Others, again, are specialists. It is not enough that a man should be versed in one department; he must be at home in all: in Botany, Zoology, Comparative Anatomy, Biology, Geology, and Palaeontology. He must be able to survey the whole field. Fourthly, and mainly, naturalists are generally lamentably deficient in philosophical culture and in a philosophical spirit. "The immovable edifice of the true, monistic science, or what is the same thing, nat-

It is very reasonable that scientific men, in common with lawyers and physicians and other professional men, should feel themselves entitled to be heard with special deference on subjects belonging to their respective departments. This deference no one is disposed to deny to men of science. But it is to be remembered that no department of human knowledge is isolated. One runs into and overlaps another. We have abundant evidence that the devotees of natural science are not willing to confine themselves to the department of nature, in the common sense of that word. They not only speculate, but dogmatize, on the highest questions of philosophy, morality, and religion. And further, admitting the special claims to deference on the part of scientific men, other men have their rights. They have the right to judge of the consistency of the assertions of men of science and of the logic of their reasoning. They have the right to set off the testimony of one or more experts against the testimony of others; and especially they have the right to reject all speculations, hypotheses, and theories which come in conflict with well-established truths. It is ground of profound gratitude to God that He has given to the human mind intuitions which are infallible, laws of belief which men cannot disregard any more than the laws of nature, and also convictions produced by the Spirit of God which no sophistry of man can weaken. These are barriers which no man can pass without plunging into the abyss of outer darkness.

If there be any truth in the preceding remarks, then it is obvious that there can be no harmony between science and religion until the evils referred to be removed. Scientific men must come to recognize practically, and not merely in words, that there are other kinds of evidence of truth than the testimony of the senses. They must come to give due weight to the testimony of consciousness and to the intuitions of the reason and conscience. They must cease to require the deference due to established facts to be paid to their speculations and explanations. And they must treat their fellow-men with due respect. The Pharisees said to the man whose sight had been restored by Christ, "Thou wast altogether born in sin, and dost

---

ural science, can only arise through the most intimate interaction and mutual interpenetration of philosophy and observation (*Philosophie und Empirik*)" (pp. 638–41). It is only a select few, therefore, of learned and philosophical monistic materialists who are entitled to be heard on questions of the highest moment to every individual man and to human society.

thou teach us?" [John 9:34]. Men of science must not speak thus. They must not say to every objector, Thou art not scientific, and therefore hast no right to speak. The true Irenicum is for all parties to give due heed to such words as these, "If any man would be wise, let him become a fool, that he may be wise" [1 Cor. 3:18]; or these, "Be converted, and become as little children" [Matt. 18:3]; or these, "The Spirit of Truth shall guide you in all truth" [John 16:13]. We are willing to hear this called cant. Nevertheless, these latter words fell from the lips of Him who spake as never man spake.

Objections to Darwinism

So much, and it is very little, on the general question of the relation of science to religion. But what is to be thought of the special relation of Mr. Darwin's theory to the truths of natural and revealed religion? We have already seen that Darwinism includes the three elements: evolution, natural selection, and the denial of design in nature. These points, however, cannot now be considered separately.

It is conceded that a man may be an evolutionist and yet not be an atheist and may admit of design in nature. But we cannot see how the theory of evolution can be reconciled with the declarations of the Scriptures. Others may see it and be able to reconcile their allegiance to science with their allegiance to the Bible. Professor Huxley, as we have seen, pronounces the thing impossible [see p. 107, n. 63]. As all error is antagonistic to truth, if the evolution theory be false, it must be opposed to the truths of religion so far as the two come into contact. Mr. Henslow, indeed, says Science and Religion are not antagonistic because they are in different spheres of thought. This is often said by men who do not admit that there is any thought at all in religion, [who maintain] that it is merely a matter of feeling. The fact, however, is that religion is a system of knowledge as well as a state of feeling. The truths on which all religion is founded are drawn within the domain of science: the nature of the first cause, its relation to the world, the nature of second causes, the origin of life, anthropology, including the origin, nature, and destiny of man. Religion has to fight for its life against a large class of scientific men. All attempts to prevent her exercising her right to be heard are unreasonable and vain.

It should be premised that this paper was written for the single purpose of answering the question, What is Darwinism? The discussion of the merits of the theory was not within the scope of the writer. What follows, therefore, is to be considered only in the light of a practical conclusion.

1. The first objection to the theory is its *prima facie* incredibility. That a single plant or animal should be developed from a mere cell is such a wonder that nothing but daily observation of the fact could induce any man to believe it. Let any one ask himself, suppose this fact was not thus familiar, what amount of speculation, of arguments from analogies, possibilities, and probabilities, could avail to produce conviction of its truth. But who can believe that all the plants and animals which have ever existed upon the face of the earth have been evolved from one such germ? This is Darwin's doctrine. We are aware that this apparent impossibility is evaded by the believers in spontaneous generation, who hold that such germ cells may be produced anywhere and at all times. But this is not Darwinism. Darwin wants us to believe that all living things, from the lowly violet to the giant redwoods of California, from the microscopic animalcule to the Mastodon, the Dinotherium—monsters the very description of which fills us with horror—bats with wings twenty feet in breadth, flying dragons, tortoises ten feet high and eighteen feet long, etc., etc., came one and all from the same primordial germ. This demand is the more unreasonable when we remember that these living creatures are not only so different, but are, as to plants and animals, directly opposed in their functions. The function of the plant, as biologists express it, is to produce force, that of the animal to expend it. The plant, in virtue of a power peculiar to itself which no art or skill of man can imitate, transmutes dead inorganic matter into organic matter suited to the sustenance of animal life, and without which animals cannot live. The gulf, therefore, between the plant and animal would seem to be impassable.

Further, the variations by which the change of species is effected are so trifling as often to be imperceptible, and their accumulation of them so slow as to evade notice—the time requisite to accomplish any marked change must be counted by millions or milliards of years. Here is another demand on our credulity. The apex is reached when we are told that all these transmutations are effected by chance, that is, without purpose or intention. Taking all these things into consideration, we think it may with moderation be said that a more absolutely incredible theory was never propounded for acceptance among men.

2. There is no pretence that the theory can be proved. Mr. Darwin does not pretend to prove it. He admits that all the facts in the case can be accounted for on the assumption of divine purpose and control. All that he claims for his theory is that it is possible. His mode of arguing is that if we suppose this and that, then it may have happened thus and so. Amiable and attractive as the man presents himself in his writings, it rouses indignation, in one class at least of his readers, to see him by such a mode of arguing reaching conclusions which are subversive of the fundamental truths of religion.

3. Another fact cannot fail to attract attention. When the theory of evolution was propounded in 1844 in the *Vestiges of Creation*, it was universally rejected; when proposed by Mr. Darwin less than twenty years afterward, it was received with acclamation. Why is this? The facts are now what they were then. They were as well known then as they are now. The theory, so far as evolution is concerned, was then just what it is now. How then is it that what was scientifically false in 1844 is scientifically true in 1864? When a drama is introduced in a theatre and universally condemned, and a little while afterward, with a little change in the scenery, it is received with rapturous applause, the natural conclusion is that the change is in the audience and not in the drama.

There is only one cause for the fact referred to that we can think of. The *Vestiges of Creation* did not expressly or effectually exclude design. Darwin does. This is a reason assigned by the most zealous advocates of his theory for their adoption of it. This is the reason given by Büchner, by Haeckel, and by Vogt. It is assigned also in express terms by Strauss, the announcement of whose death has diffused a feeling of sadness over all who were acquainted with his antecedents. In his last work, *The Old Faith and the New*, he admits "that Darwin's doctrine is a mere hypothesis; that it leaves the main points unexplained (*Die Haupt- und Cardinal-punkte noch unerklärt sind*); nevertheless, as he has shown how miracles may be excluded, he is to be applauded as one of the greatest benefactors of the human race" (p. 177). By *Wunder*, or miracle, Strauss means any event for which natural causes are insufficient to account. "We philosophers and critical theologians," he says, "have spoken well when we decreed the abolition of miracles; but our decree (*Machtspruch*) remained without effect, because we could not show them to be unnecessary, inasmuch as we were unable to indicate any natural force to take their place. Darwin has provided or indicated this natural force, this process of nature; he has opened the door through which a happier posterity may eject miracles forever." Then follows the sentence just quoted, "He who knows what hangs on miracle, will applaud Darwin as one of the greatest benefactors of the human race."

With Strauss and others of his class, miracles and design are identical, because one as well as the other assumes supernatural agency. He quotes Helmholtz,[100] who says, "Darwin's theory [is] that adaptation in the formation of organisms may arise without the intervention of intelligence, by the blind operation of natural law"; and then adds,

> As Helmholtz distinguishes the English naturalist as the man who has banished design from nature, so we have praised him as the man who has done away with miracles. Both mean the same thing.[101] Design is the miracle-worker in nature, which has put the world upside down; or as Spinoza says, has placed the last first, the effect for the cause, and thus destroyed the very idea of nature. Design in nature, especially in the department of living organisms, has ever been appealed to by those who desire to prove that the world is not self-evolved, but the work of an intelligent Creator. (p. 211)

On page 175, [Strauss] refers to those who ridicule Darwin, and yet are so far under the influence of the spirit of the age as to deny miracles or the intervention of the Creator in the course of nature, and says: "Very well; how do they account for the origin of man, and in general the development of the organic out of the inorganic? Would they assume that the original man as such, no matter how rough and unformed, but still a man, sprang immediately out of the inorganic, out of the sea or the slime of the Nile? They would hardly venture to say that; then they must know that there is only the choice between miracle, the divine hand of the Creator, and Darwin." What

---

100. [Hermann von Helmholtz (1821–94), chair of physics at the University of Berlin, rejected any explanation of life processes that appealed to nonphysical forces.]
101. This short but significant sentence is omitted in the excellent translation of Strauss's book, by Mathilde Blind. [New York, 1873.]

an alternative—the Creator or Darwin! In this, however, Strauss is right. To banish design from nature, as is done by Darwin's theory, is, in the language of the Rev. Walter Mitchell, virtually "to dethrone the Creator."

Ludwig Weis,[102] M.D., of Darmstadt, says it is at present "the mode" in Germany (and of course in a measure here) to glorify Buddhism. Strauss, he adds, says, "Nature knows itself in man, and in that he expresses the thought which all Idealism and all Materialism make the grand end. To the same effect it is said, 'In Man the All comprehends itself as conscious being (comes to self-consciousness); or, in Man the absolute knowledge (*Wissen,* the act of knowing) appears in the limits of personality.' This was the doctrine of the Buddhist and of the ancient Chinese." Thus, as Dr. Weis says, "in the nineteenth century of the Christian era, philosophers and scientists have reached the point where the Chinese were two thousand years ago."

The only way that is apparent for accounting for evolution being rejected in 1844, and for its becoming a popular doctrine in 1866, is that it happens to suit a prevailing state of mind. It is a fact, so far as our limited knowledge extends, that no one is willing to acknowledge himself not simply an evolutionist, but an evolutionist of the Darwinian school, who is not either a materialist by profession, or a disciple of Herbert Spencer, or an advocate of the philosophy of Hume.

There is another significant fact which goes to prove that the denial of design, which is the "creative idea" of Darwinism, is the main cause of its popularity and success. Professor Owen, England's greatest naturalist, is a derivationist. Derivation and evolution are convertible terms. Both include the denial that species are primordial or have each a different origin, and both imply that one species is formed out of another and simpler form. Professor Owen, however, although a derivationist or evolutionist is a very strenuous anti-Darwinian. He differs from Darwin as to two points. First, as to Natural Selection or the Survival of the Fittest. He says that is inconsistent with facts and utterly insufficient to account for the origin of species. He refers the origin of species to an inherent tendency to change impressed on them from the beginning. And second, he

---

102. [The editors have found no information on Ludwig Weis.]

admits design. He denies that the succession and origin of species are due to chance, and expresses his belief in the constant operation of creative power in the formation of species from the varied descendants of more generalized forms.[103] He believes "that all living things have been produced by such law (of variation) in time, their position and uses in the world having been preordained by the Creator."[104] Professor Owen says he has taught the doctrine of derivation (evolution) for thirty years, but it attracted little attention. As soon, however, as Darwin leaves out design, we have a prairie-fire. A prairie-fire, happily, does not continue very long; and while it lasts, it burns up little else than stubble.

4. All the evidence we have in favor of the fixedness of species is, of course, evidence not only against Darwinism, but against evolution in all its forms. It would seem idle to discuss the question of the mutability of species, until satisfied what species is. This, unhappily, is a question which it is exceedingly difficult to answer. Not only do the definitions given by scientific men differ almost indefinitely, but there is endless diversity in classification. Think of four hundred and eighty species of humming-birds. Haeckel says that one naturalist makes ten, another forty, another two hundred, and another one species of a certain fossil; and we have just heard that Agassiz had collected eight hundred species of the same fossil animal. Haeckel also says (p. 246) that there are no two zoologists or any two botanists who agree altogether in their classification. Mr. Darwin says, "No clear line of demarcation has yet been drawn between species and sub-species, and varieties" (p. 61). It is absolutely necessary, therefore, that a distinction should be made between artificial and natural species.

No man asserts the immutability of all those varieties of plants and animals which naturalists, for the convenience of classification, may call distinct species. Haeckel, for example, gives a list of twelve species of man. So any one may make fifty species of dogs or of horses. This is a mere artificial distinction which amounts to nothing. There is far greater difference between a pouter and a carrier pigeon than between a Caucasian and a Mongolian. To call the former varieties of the same species, and the latter distinct species, is altogether arbitrary.

---

103. *The Fallacies of Darwinism,* by C. R. Bree, M.D., p. 308.
104. *The Fallacies of Darwinism,* p. 305.

Nevertheless, notwithstanding the arbitrary classifications of naturalists, it remains true that there are what Professor Dana calls "units" of the organic world. "When individuals multiply from generation to generation, it is but a repetition of the primordial type-idea, and the true notion of the species is not in the resulting group, but in the idea or potential element which is the basis of every individual of the group."[105] Dr. Morton's[106] definition of species as "primordial organic forms" agrees with that given by Professor Dana, and both agree with the Bible, which says that God created plants and animals each after its kind. A primordial form is a form which was not evolved out of some other form, but which began to be in the form—subject to such varieties as we see in the dog, horse, and man—in which it continued during the whole period of its existence.

The criteria of these primordial forms or species of nature are: (1) Morphological. Animals, however, may approach very nearly in their structure and yet belong to different species. It is only when the peculiarities of structure are indicative of specialty of design that they form a safe ground of classification. If the teeth of one animal are formed to fit it to feed on flesh, and those of another to fit it to feed on plants; if one has webbed feet and another not; then, in all such cases, difference of structure proves difference of kind. (2) Physiological; that is, the internal nature, indicated by habits and instincts, furnishes another safe criterion. (3) Permanent fecundity. The progeny of the same species reproduce their kind from generation to generation; the progeny of different species, although nearly allied, do not. It is a fixed law of nature that species never can be annihilated, except by all the individuals included in them dying out, and that new species cannot be produced. Every true species is primordial. It is this fact, that is, that no variety with the essential characteristics of species has ever been produced, that forces, as we saw above, Professor Huxley to pronounce Mr. Darwin's doctrine to

105. *Bibliotheca Sacra*, 1857, p. 861. [James Dwight Dana (1813–95), noted American naturalist, held to extreme uniformitarianism in geology but, out of a desire to harmonize science and religion, to catastrophism in biology. Dana came to accept a form of evolution guided by God.]

106. [Samuel George Morton (1799–1851), anatomist and student of anthropology, founded the science of invertebrate paleontology in the United States. Morton was convinced that the human races were distinguished by the shape and capacity of their crania as well as by their skin color.]

be an unproved hypothesis [see pp. 102–4]. Species continue; varieties, if let alone, always revert to the normal type. It requires the skill and constant attention of man to keep them distinct.

Now that there are such forms in nature is proved not only from the testimony of the great body of the most distinguished naturalists, but by all the facts in the case.

First, the fact that such species are known to have existed unchanged through what geologists consider almost immeasurable periods of time. Palaeontologists tell us that Trilobites abounded from the primordial age down to the Carboniferous period, that is, as they suppose, through millions of years.[107] More wonderful still, the little animals whose remains constitute the chalk formations which are spread over large areas of country, and are sometimes a hundred feet thick, are now at work at the bottom of the Atlantic. Principal Dawson tells us, with regard to Mollusks existing in a sub-fossil state in the Post-Pliocene clays of Canada, that "after carefully studying about two hundred species, and of some of these, many hundreds of specimens, I have arrived at the conclusion that they are absolutely unchanged. . . . Here again we have an absolute refusal, on the part of all these animals, to admit that they are derived, or have tended to sport into new species."[108] On the previous page he says, "Pictet[109] catalogues ninety-eight species of mammals which inhabited Europe in the Post-glacial period. Of these fifty-seven still exist unchanged, and the remainder have disappeared. Not one can be shown to have been modified into a new form, though some of them have been obliged, by changes of temperature and other conditions, to remove into distant and now widely separated regions."

A second fact which attests the primordial character and fixedness of species is that every species as it first appears is not in a transition state between one form and another, but in the perfection of its kind. Science has indeed discovered an ascending order in creation, which agrees marvellously with that given in the book of Gen-

---

107. [As Hodge observes below, note 110, terms for geological ages were flexible. With reference to the terms used here and in the following paragraphs, the Carboniferous period was ancient (just after the Cambrian), while the Pliocene and Post-Pliocene were fairly recent. The Silurian was almost as ancient as the Cambrian.]

108. *The Story of Earth and Man*, p. 358.

109. [This is probably François Pictet de la Rive (1809–72), a zoologist and paleontologist.]

esis: first, vegetable productions; then the moving creatures in the sea; then terrestrial animals; and finally man. Naturalists who utterly reject the Scriptures as a divine revelation speak with the highest admiration of the Mosaic account of the creation as compared with any other cosmogony of the ancient world. While there is in general an ascending series in these living forms, each was perfect in its kind.

Agassiz says that fishes existed contemporaneously with species of all the invertebrate sub-kingdoms in the Taconic or sub-Cambrian strata. This is the extreme limit of known geological strata in which life is found to have existed. As the evolution of one species out of another requires, according to Darwin, millions of years, it is out of the question to trace these animals beyond the strata in which their remains are now found. Yet "crabs or lobsters, worms, cuttle-fish, snails, jelly-fish, star-fish, oysters, the polyps lived contemporaneously with the first known vertebrate animals that ever came into being—all as clearly defined by unmistakable ordinal or special characters as they are at the present moment."[110]

The foot of the horse is considered by zoologists as "one of the most beautiful contrivances in nature." The remains of this animal found in what is called the Pliocene period show the foot to have been as perfect then as it is now.

Mr. Wallace says that man has existed on the earth a hundred thousand years, and that it is probable that he existed four hundred thousand years ago. Of course we do not believe this. We have little faith in the chronology of science.[111] It gives no sure data for the calculation of time, hence we find them differing from four thousand to four hundred thousand years as to the time required for certain formations. The most trustworthy geologists teach that all that is

---

110. Dr. Bree, p. 275. We presume geologists differ in the terms which they use to designate strata. Agassiz calls the oldest containing fossils the sub-Cambrian. Principal Dawson calls the oldest the Laurentian and places the first vertebrates in the Silurian. This is of no moment as to the argument. The important fact is that each species is distinct as soon as it appears, and that many have remained to the present time.

111. [William Henry Green, Hodge's colleague in Old Testament at Princeton Seminary, later published an influential article on the genealogies of Genesis which convinced B. B. Warfield that the biblical record was compatible with a very old age for human beings—Green, "Primeval Chronology," *Bibliotheca Sacra* 47 (April 1890): 285–303.]

known of the antiquity of man falls within the limits of Biblical chronology. The further, however, Darwinians push back the origin of man, the stronger, as against them, becomes the argument for the immutability of species. The earliest remains of man show that at his first appearance he was in perfection. The oldest known human skull is that called the "Engis," because found in the cave of Engis in Belgium.[112] Of this skull Professor Huxley says it may have belonged to an individual of one of the existing races of men. Principal Dawson, who has a cast of it on the same shelf with the skulls of some Algonquin Indians, says it might be taken for the skull of an American Indian. Indeed, Dawson seems to think that these fossil human remains go to show that the earliest men were better developed than any of the extant races.

Thirdly, the historical evidence accessible all goes to prove the immutability of species. The earliest historical records and the oldest monuments prove that all extant animals [are now] what they [were] thousands of years ago.

Fourthly, the fact that hybrids cannot be perpetuated, that no device of man can produce a new species, is proof that God has fixed limits which cannot be passed. This Huxley himself admits to be an insuperable objection. So long as it exists, he says, Darwin's doctrine must be content to remain a hypothesis; it cannot pretend to the dignity of a theory.

Another fact of like import is that varieties artificially produced, if let alone, uniformly revert to the simple typical form. It is only by the utmost care they can be kept distinct. All the highly prized varieties of horses, cattle, sheep, pigeons, etc., without human control would be merged each class into one, with only the slight differences occasioned by diversities of climate and other external conditions. If in the sight of man it is important that the words of a book should be kept distinct, it is equally evident that in the sight of God it is no less important that the "units of nature" should not be mixed in inextricable and indistinguishable confusion.[113]

---

112. [In 1832 P. C. Schmerling found old human remains, including a skull, near Liège, Belgium, in what was called the "Engis cave." The nature of the human being was hotly contested, because the skull had both a low forehead (suggesting a primitive condition) and a large cranial capacity (suggesting advanced development).]

113. [Almost all of the nineteenth-century speculation on inheritance—whether Darwin's and Huxley's or Hodge's and his authorities—would be made obsolete by the end of the century with the rise of Mendelian genetics.]

Fifthly, the sudden appearance of new kinds of animals is another fact which Palaeontologists urge against the doctrine of evolution. According to the view of geologists great changes have, at remote periods, occurred in the state of the earth. Continents have been submerged and the bottom of the sea raised above the surface of the waters. Corresponding changes have occurred in the state of the atmosphere surrounding the globe and in the temperature of the earth. Accompanying or following these revolutions new classes of plants and animals appear, adapted to the new condition of the earth's surface. Whence do they come? They have, as Dawson expresses it, neither fathers nor mothers. Nothing precedes them from which they could be derived, and nothing of the same kind follows them. They live through their appointed period and then, in a multitude of cases, finally disappear, and are in their turn followed by new orders or kinds. In other words, the links or connecting forms of this assumed regular succession or derivation are not to be found. This fact is so patent that Hugh Miller,[114] when arguing against the doctrine of evolution as proposed in the *Vestiges of Creation*, says that the record in the rocks seems to have been written for the very purpose of proving that such evolution is impossible.

We have the explicit testimony of Agassiz as a Palaeontologist that the facts of geology contradict the theory of the transmutation of species. This testimony has been repeatedly given and in various forms. In the last production of his pen he says:

> As a Palaeontologist I have from the beginning stood aloof from this new theory of transmutation, now so widely admitted by the scientific world. Its doctrines, in fact, contradict what the animal forms buried in the rocky strata of our earth tell us of their own introduction and succession upon the surface of the globe. . . . Let us look now at the earliest vertebrates, as known and recorded in geological surveys. They should, of course, if there is any truth in the transmutation theory, correspond with the lowest in rank or standing. What then are the earliest known vertebrates? They are Selachians (sharks and their allies) and Ganoids (garpikes and the like), the highest of all living fishes, structurally speaking.

---

114. [Hugh Miller (1802–56), prominent Scottish stonemason-geologist, held that the fossil record confirmed the creation sequences recorded in the Mosaic narrative. Miller was disturbed by *Vestiges* because it obliterated the sharp line between animals and humanity.]

He closes the article from which these quotations are taken with the assertion "that there is no evidence of a direct descent of later from earlier species in the geological succession of animals."[115]

It will be observed that Agassiz is quoted, not as to matters of theory, but as to matters of fact. The only answer which evolutionists can make to this argument is the imperfection of the geological record. When asked, Where are the immediate predecessors of these new species? they answer, They have disappeared or have not yet been found. When asked, Where are their immediate successors? the answer again is, They have disappeared.[116] This is an objection which Mr. Darwin, with his usual candor, virtually admits to be unanswerable. We have already seen that he says, "Every one will admit that the geological record is imperfect; but very few can believe that it is so very imperfect as my theory demands."

Such are some of the grounds on which geologists and palaeontologists of the highest rank assert that the theory of evolution has not the slightest scientific basis, and they support their assertion with an amount of evidence of which the above items are a miserable pittance.

The Grand and Final Objection

Sixthly, there is another consideration of decisive importance. Strauss says there are three things which have been stumbling-blocks in the way of science. First, the origin of life; second, the origin of consciousness; third, the origin of reason. These are equivalent to the gaps which, Principal Dawson says, exist in the theory of evolution. He states them thus: (1) That between dead and living matter. (2) That between vegetable and animal life. "These are necessarily the converse of each other: the one deoxidizes and accumulates, the other oxidizes and expends." (3) That "between any species of plant or animal, and any other species. It was this gap, and this only, which Darwin undertook to fill up by his great work on the origin of species, but, notwithstanding the immense

---

115. *Atlantic Monthly*, January, 1874.

116. We have heard a story of a gentleman who gave an artist a commission for a historical painting, and suggested as the subject the passage of the Israelites over the Red Sea. In due time he was informed that his picture was finished, and was shown by the artist a large canvas painted red. "What is that?" he asked. "Why," says the artist, "that is the Red Sea." "But where are the Israelites?" "Oh, they have passed over." "And where are the Egyptians?" "They are under the sea."

amount of material thus expended, it yawns as wide as ever, since it must be admitted that no case has been ascertained in which an individual of one species has transgressed the limits between it and another species." (4) "Another gap is between the nature of the animal and the self-conscious, reasoning, and moral nature of man" (pp. 325–28).

First, as to the gap between death and life, this is what Dr. Stirling calls the "gulf of all gulfs, which Mr. Huxley's protoplasm is as powerless to efface as any other material expedient that has ever been suggested."[117] This gulf Mr. Darwin does not attempt to bridge over. He admits that life owes its origin to the act of the Creator. This, however, the most prominent of the advocates of Darwinism say, is giving up the whole controversy. If you admit the intervention of creative power at one point, you may as well admit it in any other. If life owes its origin to creative power, why not species? If the stupendous miracle of creation be admitted, there is no show of reason for denying supernatural intervention in the operations of nature.

Most Darwinians attempt to pass this gulf on the imaginary bridge of spontaneous generation. In other words, they say there is no gulf there. The molecules of matter in one combination may as well exhibit the phenomena of life as in other combinations [they exhibit] any other kind of phenomena. The distinguished Sir William Thomson cannot trust himself to that bridge. "Dead matter," he says, "cannot become living matter without coming under the influence of matter previously alive. This seems to me as sure a teaching of science as the law of gravitation. . . . I am ready to adopt, as an article of scientific faith, true through all space and through all time, that life proceeds from life, and nothing but life."[118] He

---

117. *As Regards Protoplasm in Relation to Professor Huxley's Essay on the Physical Basis of Life.* By James H. Stirling. See also *Physiological Anatomy and Physiology of Man,* by L. S. Beale; also, *The Mystery of Life in Reply to Dr. Gull's Attack on the Theory of Vitality.* By L. S. Beale, M.D., 1871. [Stirling (1820–1909) was a Scottish philosopher who thought he could employ both Immanuel Kant and Georg Hegel to develop a modern form of scientific natural theology.]

118. The address delivered by Sir William Thomson as President of the British Association at its meeting in Edinburgh, 1871. [Thomson (1824–1907), later Lord Kelvin, was the leading physicist in nineteenth-century Britain. He was an opponent of Charles Lyell's uniformitarian approach to geology, and his essays in the early 1860s on the age of the earth troubled Darwin, for they did not allow the expanse of time that his theory of natural selection required.]

refers the origin of life on this earth to falling meteors, which bring with them from other planets the germs of living organisms; and from those germs all the plants and animals with which our world is now covered have been derived. Principal Dawson thinks that this was intended as irony. But the whole tone of the address, and specially of the closing portion of it, in which this idea is advanced, is far too serious to admit of such an explanation.

No one can read the address referred to without being impressed, and even awed, by the immensity and grandeur of the field of knowledge which falls legitimately within the domain of science. The perusal of that discourse produces a feeling of humility analogous to the sense of insignificance which every man experiences when he thinks of himself as a speck on the surface of the earth, which itself is but a speck in the immensity of the universe. And when a man of mere ordinary culture sees Sir William Thomson surveying that field with a mastery of its details and familiarity with all the recondite methods of its investigation, he feels as nothing in his presence. Yet this great man, whom we cannot help regarding with wonder, is so carried away by the spirit of his class as to say, "Science is bound, by the everlasting law of honor, to face fearlessly every problem which can fairly be brought before it. If a probable solution, consistent with the ordinary course of nature, can be found, we must not invoke an abnormal act of Creative Power." And, therefore, instead of invoking Creative Power, he accounts for the origin of life on earth by falling meteors. How he accounts for its origin in the places whence the meteors came, he does not say. Yet Sir William Thomson believes in Creative Power, and in a subsequent page we shall quote his explicit repudiation of the atheistic element in the Darwinian theory.

Strauss quotes Dubois-Reymond,[119] a distinguished naturalist, as teaching that the first of these great problems, viz. the origin of life, admits of explanation on scientific (*i.e.*, in his sense, materialistic) principles; and even the third, viz. the origin of reason; but the second, or the origin of consciousness, he says "is perfectly inscrutable." Dubois-Reymond holds that the most accurate knowledge of the essential organism reveals to us only matter in motion; but between this material movement and my feeling pain or pleasure,

---

119. [Emil Dubois-Reymond (1818–96), German physiologist and philosopher of science, held that metaphysical questions are in principle unsolvable.]

experiencing a sweet taste, seeing red, with the conclusion "therefore I exist," there is a profound gulf; and it "remains utterly and forever inconceivable why to a number of atoms of carbon, hydrogen, etc., it should not be a matter of indifference how they lie or how they move; nor can we in any wise tell how consciousness should result from their concurrent action." "Whether," adds Strauss, "these *Verba Magistri* [words of the master] are indeed the last word on the subject, time only can tell."[120]

But if it is inconceivable, not to say absurd, that sense-consciousness should consist in the motion of molecules of matter, or be a function of such molecules, it can hardly be less absurd to account for thought, conscience, and religious feeling and belief on any such hypothesis. It may be said that Mr. Darwin is not responsible for these extreme opinions. That is very true. Mr. Darwin is not a Monist, for in admitting creation he admits a dualism as between God and the world. Neither is he a Materialist, inasmuch as he assumes a supernatural origin for the infinitesimal modicum of life and intelligence in the primordial animalcule from which, without divine purpose or agency, all living things in the whole history of our earth have descended. All the innumerable varieties of plants, all the countless forms of animals with all their instincts and faculties, all the varieties of men with their intellectual endowments and their moral and religious nature, have, according to Darwin, been evolved by the agency of the blind, unconscious laws of nature. This infinitesimal spark of supernaturalism in Mr. Darwin's theory would inevitably have gone out of itself, had it not been rudely and contemptuously trodden out by his bolder and more logical successors.

The grand and fatal objection to Darwinism is this exclusion of design in the origin of species or the production of living organisms. By design is meant the intelligent and voluntary selection of an end, and the intelligent and voluntary choice, application, and control of means appropriate to the accomplishment of that end. That design therefore implies intelligence is involved in its very nature. No man can perceive this adaptation of means to the accomplishment of a preconceived end without experiencing an irresistible conviction that it is the work of mind. No man does doubt it, and no man can

---

120. *The Old Faith and the New*. Prefatory Postscript, xxi.

doubt it. Darwin does not deny it. Haeckel does not deny it. No Darwinian denies it. What they do is to deny that there is any design in nature. It is merely apparent, as when the wind of the Bay of Biscay, as Huxley says, "selects the right kind of sand and spreads it in heaps upon the plains." But in thus denying design in nature, these writers array against themselves the intuitive perceptions and irresistible convictions of all mankind—a barrier which no man has ever been able to surmount. Sir William Thomson, in the address already referred to, says:

> I feel profoundly convinced that the argument of design has been greatly too much lost sight of in recent zoological speculations. Reaction against the frivolities of teleology, such as are to be found, not rarely, in the notes of the learned commentators on "Paley's Natural Theology," has, I believe, had a temporary effect of turning attention from the solid irrefragable argument so well put forward in that excellent old book. But overpowering proof of intelligence and benevolent design lie all around us, and if ever perplexities, whether metaphysical or scientific, turn us away from them for a time, they come back upon us with irresistible force, showing to us through nature the influence of a free will, and teaching us that all living beings depend upon one ever-acting Creator and Ruler.

It is impossible for even Mr. Darwin, inconsistent as it is with his whole theory, to deny all design in the constitution of nature. What is his law of heredity? Why should like beget like? Take two germ cells, one of a plant, another of an animal; no man by microscope or by chemical analysis or by the magic power of the spectroscope can detect the slightest difference between them, yet the one infallibly develops into a plant and the other into an animal. Take the germ of a fish and of a bird, and they are equally indistinguishable; yet the one always under all conditions develops into a fish and the other into a bird. Why is this? There is no physical force, whether light, heat, electricity, or anything else, which makes the slightest approximation to accounting for that fact. To say, as Stuart Mill[121] would say, that it is an ultimate fact, and needs no explanation, is to say that there may be an effect without an adequate cause. The venerable

---

121. [John Stuart Mill (1806–73), influential promoter of utilitarian ethics, attacked especially those philosophers, like Sir William Hamilton, who sought a transcendental dimension to life.]

K. E. Von Baer,[122] the first naturalist in Russia, of whom Agassiz speaks in terms of such affectionate veneration in the *Atlantic Monthly* for January, 1874, has written a volume dated Dorpat, 1873, and entitled *Zum Streit über den Darwinismus* [*The Battle over Darwinism*]. In that volume, as we learn from a German periodical, the author says:

> The Darwinians lay great stress on heredity; but what is the law of heredity but a determination of something future? Is it not in its nature in the highest degree teleological? Indeed, is not the whole faculty of reproduction intended to introduce a new life-process? When a man looks at a dissected insect and examines its strings of eggs, and asks, Whence are they? the naturalist of our day has no answer to give, but that they were of necessity gradually produced by the changes in matter. When it is further asked, Why are they there? is it wrong to say, It is *in order that* when the eggs are mature and fertilized, new individuals of the same form should be produced?

It is further to be considered that there are innumerable cases of contrivance, or evidence of design in nature, to which the principle of natural selection, or the purposeless changes effected by unconscious force, cannot apply, as for example, the distinction of sex, with all that is therein involved. But passing by such cases, it may be asked, what would it avail to get rid of design in the vegetable and animal kingdom, while the whole universe is full of it? That this ordered cosmos is not from necessity or chance is almost a self-evident fact. Not one man in a million of those who ever heard of God either does doubt or can doubt it. Besides, how are the cosmical relations of light, heat, electricity, to the constituent parts of the universe and especially, so far as this earth is concerned, to vegetable and animal life, to be accounted for? Is this all chance work? Is it by chance that light and heat cause plants to carry on their wonderful operations, transmuting the inorganic into the organic, dead matter into living and life-sustaining matter? Is it without a purpose that water instead of contracting expands at the freezing point—a fact to which is due that the earth north of the tropic is habitable for man or beast? It is

---

122. [Karl Ernst Von Baer (1792–1876), while a believer in the transformation of species, held firmly to the notion of design in nature. A student of embryology and anthropology, he also opposed Darwin's idea that all organisms evolved from a single or a few progenitors.]

no answer to this question to say that a few other substances have the same peculiarity when no good end, that we can see, is thereby accomplished. No man is so foolish, because he cannot tell what the spleen was made for, as to deny that his eye was intended to enable him to see. It is, however, useless to dwell upon this subject. If a man denies that there is design in nature, he can with quite as good reason deny that there is any design in any or in all the works ever executed by man.

The conclusion of the whole matter is that the denial of design in nature is virtually the denial of God. Mr. Darwin's theory does deny all design in nature; therefore, his theory is virtually atheistical—his theory, not he himself. He believes in a Creator. But when that Creator, millions on millions of ages ago, did something—called matter and a living germ into existence—and then abandoned the universe to itself to be controlled by chance and necessity, without any purpose on his part as to the result, or any intervention or guidance, then He is virtually consigned, so far as we are concerned, to non-existence.

It has already been said that the most extreme of Mr. Darwin's admirers adopt and laud his theory for the special reason that it banishes God from the world, that it enables them to account for design without referring it to the purpose or agency of God. This is done expressly by Büchner, Haeckel, Vogt, and Strauss. The opponents of Darwinism direct their objections principally against this element of the doctrine. This, as was stated by Rev. Dr. Peabody, was the main ground of the earnest opposition of Agassiz to the theory. America's great botanist, Dr. Asa Gray, avows himself an evolutionist, but he is not a Darwinian. Of that point we have the clearest possible proof. Mr. Darwin, after explicitly denying that the variations which have resulted in "the formation of the most perfectly adapted animals in the world, man included, were intentionally and specially guided," adds: "However much we may wish it, we can hardly follow Professor Asa Gray in his belief 'that variation has been led along certain beneficial lines' like a stream 'along definite and useful lines of irrigation.'"[123] If Mr. Darwin does not agree with Dr. Gray, Dr. Gray does not agree with Mr. Darwin. It is as to the exclusion of design from the operations of nature that our American differs from the English

---

123. *Variation of Animals and Plants under Domestication.* [New York, 1868, vol. 2, pp. 515–16.]

naturalist. This is the vital point. The denial of final causes is the formative idea of Darwin's theory, and therefore no teleologist can be a Darwinian.

Dr. Gray quotes from another writer the sentence, "It is a singular fact, that when we can find how anything is done, our first conclusion seems to be that God did not do it"; and then adds,

> I agree with the writer that this first conclusion is premature and unworthy; I will add, deplorable. Through what faults of dogmatism on the one hand, and skepticism on the other, it came to be so thought, we need not here consider. Let us hope, and I confidently expect, that it is not to last; that the religious faith which survived without a shock the notion of the fixedness of the earth itself, may equally outlast the notion of the absolute fixedness of the species which inhabit it; that in the future, even more than in the past, faith in an *order*, which is the basis of science, will not—as it cannot reasonably—be dissevered from faith in an *Ordainer*, which is the basis of religion.[124]

We thank God for that sentence. It is the concluding sentence of Dr. Gray's address as ex-President of the American Association for the Advancement of Science, delivered August, 1872.

Dr. Gray goes further. He says, "The proposition that the things and events in nature were not designed to be so, if logically carried out, is doubtless tantamount to atheism." Again, "To us, a fortuitous Cosmos is simply inconceivable. The alternative is a designed Cosmos.... If Mr. Darwin believes that the events which he supposes to have occurred and the results we behold around us were undirected and undesigned; or if the physicist believes that the natural forces to which he refers phenomena are uncaused and undirected, no argument is needed to show that such belief is atheistic."[125]

We have thus arrived at the answer to our question, What is Darwinism? It is Atheism. This does not mean, as before said, that Mr. Darwin himself and all who adopt his views are atheists; but it means

---

124. *Proceedings of the American Association for the Advancement of Science.* [Cambridge, 1873, p. 20.]
125. The *Atlantic Monthly* for October 1860. The three articles in the July, August, and October numbers of the *Atlantic* on the subject have been reprinted with the name of Dr. Asa Gray as their author.

that his theory is atheistic, that the exclusion of design from nature is, as Dr. Gray says, tantamount to atheism.

Among the last words of Strauss were these:

> We demand for our universe the same piety which the devout man of old demanded for his God.... In the enormous machine of the universe, amid the incessant whirl and hiss of its jagged iron wheels, amid the deafening crash of its ponderous stamps and hammers, in the midst of this whole terrific commotion, man, a helpless and defenceless creature, finds himself placed, not secure for a moment that on an imprudent motion a wheel may not seize and rend him, or a hammer crush him to a powder. This sense of abandonment is at first something awful.[126]

Among the last words of Paul were these:

> I know whom I have believed, and am persuaded that He is able to keep that which I have committed unto Him against that day.... The time of my departure is at hand. I have fought a good fight, I have finished my course, I have kept the faith: henceforth there is laid up for me a crown of righteousness, which the Lord, the righteous judge, shall give me at that day: and not to me only, but unto all them also that love his appearing. [2 Tim. 1:12; 4:6–8]

---

126. Strauss says that as he has arrived at the conclusion that there is no personal God, and no life after death, it would seem to follow that the question, Have we still a religion? "must be answered in the negative." But as he makes the essence of religion to consist in a sense of dependence, and as he felt himself to be helpless in the midst of this whirling universe, he had that much religion left.

Asa Gray's Review of *What Is Darwinism?*

In the Introduction we note the importance of Asa Gray's review of *What Is Darwinism?* both for the merits of his case against Hodge and for more general complexities surrounding the definition of "Darwinism." Gray's review is particularly important because he was both the leading American promoter of Darwin's ideas and a fairly conservative, relatively traditional Christian who remained an active Congregationalist even at Unitarian Harvard. At the time of his retirement in 1873, Gray was asked his religious views and how they compared with Darwin. In response he replied that he was "a humble member of the Christian church, & as for orthodoxy, I receive & profess, *ex animo,* the Nicene creed for myself, though with no call to deny the Christian name to those who receive less." On the comparative question Gray concluded, "I dare say I am much more orthodox than Mr. Darwin; also that he is about as far from being an atheist as I am."[1]

Gray's review was by far the most discriminating to have appeared in Hodge's lifetime. Whether or not he sustained his arguments—that Darwin's natural selection really can be profoundly teleological and that Hodge was more interested in promoting an interventionist view of God's providence than he was in unpacking the character of Darwinism—he greatly advanced the awareness of how important it was to say with utmost clarity exactly what was at stake in debating the character of Darwin's ideas.

For his part, Hodge took Gray's criticisms in stride. According to his grandson William Berryman Scott, who had helped read proof for *What Is Darwinism?* Hodge made this satisfied comment after read-

1. A. Hunter Dupree, *Asa Gray* (Cambridge, Mass.: Harvard University Press, 1959), 358–59.

ing the review: "Gray admits that I have stated Darwin's position fairly, and I don't care about the rest."[2]

Gray's review was later reprinted in a collection of his essays, *Darwiniana: Essays and Reviews Pertaining to Darwinism* (New York: Appleton, 1876), which was prepared for the press in part by George Frederick Wright, the leading proponent in his day for the idea that Darwin's theory, properly understood, comported well with traditional Calvinistic orthodoxy.[3] Gray's *Darwiniana* is available in a modern edition, edited by A. Hunter Dupree, from Harvard University Press (1963). The material that follows is taken from the original printing in *The Nation* (28 May 1874, pp. 348–51), whose orthography and punctuation were later revised very slightly for publication in Gray's book.[4] (We have omitted Gray's brief comments on three other books that were included with his extensive consideration of *What Is Darwinism?*)

The question which Dr. Hodge asks he promptly and decisively answers: "What is Darwinism? It is atheism."

Leaving aside all subsidiary and incidental matters, let us consider (1) what the Darwinian doctrine is, and (2) how it is proved to be atheistic. Dr. Hodge's own statement of it cannot be very much bettered:

> His [Darwin's] work on the *Origin of Species* does not purport to be philosophical. In this aspect it is very different from the cognate works of Mr. Spencer. Darwin does not speculate on the origin of the universe, on the nature of matter, or of force. He is simply a naturalist, a careful and laborious observer, skillful in his descriptions, and singularly candid in dealing with the difficulties in the way of his peculiar doctrine. He set before himself a single problem, namely, How are the fauna and flora of our earth to be accounted for? . . . To account for the existence of matter and life, Mr. Darwin admits a Creator. This is done explicitly and repeatedly. . . . He assumes the efficiency of physical causes, s*howing no disposition to resolve them into mind-*

---

2. William Berryman Scott, *Some Memories of a Palaeontologist* (Princeton: Princeton University Press, 1939), 49.

3. See especially Ronald L. Numbers, "George Frederick Wright: From Christian Darwinist to Fundamentalist," *Isis* 79 (1988): 624–45.

4. The most significant change is that Gray did not capitalize "Nature" in the essay's first appearance.

force, or into the efficiency of the First Cause. . . . He assumes also the existence of life in the form of one or more primordial germs. . . . How all living things on earth, including the endless variety of plants, and all the diversity of animals, . . . have descended from the primordial animalcule, he thinks, may be accounted for by the operation of the following natural laws: First, the law of Heredity, or that by which like begets like. The offspring are like the parent. Second, the law of Variation, that is, while the offspring are in all essential characteristics like their immediate progenitor, they nevertheless vary, more or less within narrow limits, from their parent and from each other. Some of these variations are indifferent, some deteriorations, some improvements, that is, they are such as enable the plant or animal to exercise its functions to greater advantage. Third, the law of Over-Production. All plants and animals tend to increase in a geometrical ratio, and therefore tend to overrun enormously the means of support. If all the seeds of a plant, all the spawn of a fish, were to arrive at maturity, in a very short time the world could not contain them. Hence of necessity arises a struggle for life. Only a few of the myriads born can possibly live. Fourth, here comes in the law of Natural Selection, or the Survival of the Fittest. That is, if any individual of a given species of plant or animal happens to have a slight deviation from the normal type, favorable to its success in the struggle for life, it will survive. This variation, by the law of heredity, will be transmitted to its offspring, and by them again to theirs. Soon these favored ones gain the ascendency, and the less favored perish; and the modification becomes established in the species. After a time another and another of such favorable variations occur, with like results. Thus, very gradually, great changes of structure are introduced, and not only species, but genera, families, and orders in the vegetable and animal world are produced. [pp. 78–79]

Now, the truth or the probability of Darwin's hypothesis is not here the question, but only its congruity or incongruity with theism. We need take only one exception to this abstract of it, but that is an important one for the present investigation. It is to the sentence which we have italicized in the earlier part of Dr. Hodge's own statement of what Darwinism is. With it begins our inquiry as to how he proves the doctrine to be atheistic.

First, if we rightly apprehend it, a suggestion of atheism is infused into the premises in a negative form: Mr. Darwin shows no disposition to resolve the efficiency of physical causes into the efficiency of the First Cause. Next (on page 89) comes the positive charge that

"Mr. Darwin, although himself a theist," maintains that "the contrivances manifested in the organs of plants and animals . . . are not due to the continued cooperation and control of the divine mind, nor to the original purpose of God in the constitution of the universe." As to the negative statement, it might suffice to recall Dr. Hodge's truthful remark that Darwin "is simply a naturalist," and that "his work on the *Origin of Species* does not purport to be philosophical." In physical and physiological treatises, the most religious men rarely think it necessary to postulate the First Cause, nor are they misjudged by the omission. But surely Mr. Darwin does show the disposition which our author denies him, not only by implication in many instances, but most explicitly where one would naturally look for it, namely, at the close of the volume in question: "To my mind, it accords better with what we know of the laws impressed on matter by the Creator," etc. If that does not refer the efficiency of physical causes to the First Cause, what form of words could do so? The positive charge appears to be equally gratuitous. In both Dr. Hodge must have overlooked the beginning as well as the end of the volume which he judges so hardly. Just as mathematicians and physicists in their systems are wont to postulate the fundamental and undeniable truths they are concerned with, or what they take for such and require to be taken for granted, so Mr. Darwin postulates, upon the first page of his notable work, and in the words of Whewell and Bishop Butler, (1) The establishment by divine power of general laws according to which, rather than by insulated interpositions in each particular case, events are brought about in the material world; and (2) That by the word *natural* is meant "stated, fixed, or settled" by this same power, "since what is natural as much requires and presupposes an intelligent agent to render it so—i.e., to effect it continually or at stated times—as what is supernatural or miraculous does to effect it for once." So when Mr. Darwin makes such large and free use of "natural as antithetical to supernatural" causes, we are left in no doubt as to the ultimate source which he refers them to. Rather let us say there ought to be no doubt, unless there are other grounds for it to rest upon.

Such ground there must be, or seem to be, to justify or excuse a veteran divine and scholar like Dr. Hodge in his deduction of pure atheism from a system produced by a confessed theist and based, as we have seen, upon thoroughly orthodox fundamental conceptions. Even if we may not hope to reconcile the difference between the theologian and the naturalist, it may be well to ascertain where their real divergence begins, or ought to begin, and what it amounts to. Seemingly, it is in their proximate, not in their ultimate, principles, as Dr. Hodge insists when he declares that the whole drift of Darwinism is to prove that everything "may be accounted for by the blind operation of natural causes, without any intention, purpose, or cooperation of God" (p. 98). "Why don't he say," cries the theologian, "that the complicated organs of plants and animals are the product of the divine intelligence? If God made them, it makes no difference, so far as the question of design is concerned, how He made them: whether at once or by a process of evolution" (p. 95). But, as we have seen, Mr. Darwin does say that, and he over and over implies it when he refers the production of species "to secondary causes," and likens their origination to the origination of individuals, species being series of individuals with greater difference. It is not for the theologian to object that the power which made individual men and other animals, and all the differences which the races of mankind exhibit, through secondary causes, could not have originated congeries of more or less greatly differing individuals through the same causes.

Clearly, then, the difference between the theologian and the naturalist is not fundamental, and evolution may be as profoundly and as particularly theistic as it is increasingly probable. The taint of atheism which, in Dr. Hodge's view, leavens the whole lump is not inherent in the original grain of Darwinism—in the principles posited—but has somehow been introduced in the subsequent treatment. Possibly, when found, it may be eliminated. Perhaps there is mutual misapprehension growing out of some ambiguity in the use of terms. "Without any intention, purpose, or cooperation of God." These are sweeping and effectual words. How came they to be applied to natural selection by a divine who professes that God ordained whatsoever cometh to pass? In this wise: "The point to be proved is that it is the distinctive doctrine of Mr. Darwin that species owe their origin (1) not to the original intention of the divine mind, (2) not to special acts of creation calling new forms into existence at certain epochs, (3) not to the constant and everywhere operative efficiency of God, guiding physical causes in the production of intended effects, but (4) to the gradual accumulation of *unintended* variations of structure and instinct, securing some advantage to their subjects" (p. 92). Then Dr. Hodge adduces "Darwin's own testimony" to the purport that natural selection denotes the totality of natural causes and their interactions, physical and physiological, reproduction, variation, birth, struggle, extinction—in short, all that is going on in nature; that the variations which in this interplay are picked out for survival are *not intentionally guided;* that "nothing can be more hopeless than to attempt to explain this similarity of pattern in members of the same class, by utility or the doctrine of final causes" (which Dr. Hodge takes to be the denial of any such thing as final causes); and that the interactions and processes going on which constitute natural selection may suffice to account for the present diversity of animals and plants (primordial organisms being postulated and time enough given) with all their structures and adaptations—that is, to account for them scientifically, as science accounts for other things.

A good deal may be made of this, but does it sustain the indictment? Moreover, the counts of the indictment may be demurred to. It seems to us that only one of the three points which Darwin is said to deny is really opposed to the fourth, which he is said to maintain, except as concerns the perhaps ambiguous word *unintended*. Otherwise, the origin of species through the gradual accumulation of variations—i.e., by the addition of a series of small differences—is surely not incongruous with their origin through "the original intention of the divine mind" or through "the constant and everywhere operative efficiency of God." One or both of these Mr. Darwin (being, as Dr. Hodge says, a theist) must needs hold to in some form or other; wherefore he may be presumed to hold the fourth proposition in such wise as not really to contradict the first or the third. The proper antithesis is with the second proposition only, and the issue comes to this: Have the multitudinous forms of living creatures, past and present, been produced by as many special and independent acts of creation at very numerous epochs? Or have they originated under causes as natural as reproduction and birth, and no more so, by the variation and change of preceding into succeeding species?

Those who accept the latter alternative are evolutionists. And Dr. Hodge fairly allows that their views, although clearly wrong, may be genuinely theistic. Surely they need not become the less so by the discovery or by the conjecture of natural operations through which this diversification and continued adaptation of species to conditions is brought about. Now, Mr. Darwin thinks—and by this he is distinguished from most evolutionists—that he can assign actual natural causes adequate to the production of the present out of the preceding state of the animal and vegetable world, and so on backward—thus uniting, not indeed the beginning, but the far past with the present in one coherent system of nature. But in assigning actual natural causes and processes, and applying them to the explanation of the whole case, Mr. Darwin assumes the obligation of maintaining their general sufficiency—a task from which the numerous advocates and acceptors of evolution on the general concurrence of probabilities and its usefulness as a working hypothesis (with or without much conception of the manner how) are happily free. Having hit upon a *modus operandi* which all who understand it admit will explain something, and many that it will explain very much, it is to be expected that Mr. Darwin will make the most of it. Doubtless he is far from pretending to know all the causes and operations at work; he has already added some and restricted the range of others; he probably looks for additions to their number and new illustrations of their efficiency; but he is bound to expect them all to fall within the category of what he calls natural selection (a most expansible principle), or to be congruous with it—that is, that they shall be natural causes. Also—and this is the critical point—he is bound to maintain their sufficiency without *intervention*.

Here, at length, we reach the essential difference between Darwin, as we understand him, and Dr. Hodge. The terms which Darwin sometimes uses, and doubtless some of the ideas they represent, are not such as we should adopt or like to defend; and we may say once for all—aside though it be from the present issue—that, in our opinion, the adequacy of the assigned causes to the explanation of the phenomena has not been made out. But we do not understand him to deny "purpose, intention, or the cooperation of God" in nature. This would be as gratuitous as unphilosophical, not to say unscientific. When he speaks of this or that particular or phase in the course of events or the procession of organic forms as not intended, he seems to mean not specially and disjunctively intended and not brought about by intervention. Purpose in the whole, as we suppose, is not denied but implied. And when one considers how, under whatever view of the case, the designed and the contingent lie inextricably commingled in this world of ours, past man's disentanglement, and into what metaphysical dilemmas the attempt at unraveling them leads, we cannot greatly blame the naturalist for relegating such problems to the philosopher and the theologian. If charitable, these will place the most favorable construction upon attempts to extend and unify the operation of known secondary causes, this being the proper business of the naturalist and physicist; if wise, they will be careful not to predicate or suggest the absence of intention from what comes about by degrees through the continuous operation of physical causes, even in the organic world, lest, in their endeavor to retain a probable excess of supernaturalism in that realm of nature, they cut away the grounds for recognizing it at all in inorganic nature, and so fall into the same condemnation that some of them award to the Darwinian.

Moreover, it is not certain that Mr. Darwin would very much better his case, Dr. Hodge being judge, if he did propound some theory of the *nexus* of divine causation and natural laws, or even if he explicitly adopted the one or the other of the views which he is charged with rejecting. Either way he might meet a procrustean fate; and, although a saving amount of theism might remain, he would not be sound or comfortable. For if he predicates "the constant and everywhere operative efficiency of God," he may "lapse into the same doctrine" that the Duke of Argyll and Sir John Herschel "seem inclined to," the latter of whom is blamed for thinking "it but reasonable to regard the force of gravitation as the direct or indirect result of a consciousness or will existing somewhere," and the former for regarding "it unphilosophical 'to think or speak as if the forces of nature were either independent of, or even separate from the Creator's power'" (pp. 76–77); while if he falls back upon an "original intention of the divine mind" endowing matter with forces which he foresaw and intended should produce such results as these contrivances in nature, he is told (pp. 87–88) that this banishes God from the world and is inconsistent with obvious facts. And that because of its implying that "He never *interferes* to guide the operation of physical causes." We italicize the word, for *interference* proves to be the keynote of Dr. Hodge's system. Interference with a divinely ordained physical nature for the accomplishment of natural results! An unorthodox friend has just imparted to us, with much misgiving and solicitude lest he should be thought irreverent, his tentative hypothesis, which is, that even the Creator may be conceived to have improved with time and experience! Never before was this theory so plainly and barely put before us. We were obliged to say that in principle and by implication it was not wholly original.

But in such matters, which are far too high for us, no one is justly to be held responsible for the conclusions which another may draw from his principles or assumptions. Dr. Hodge's particular view should be gathered from his own statement of it:

> In the external world there is always and everywhere indisputable evidence of the activity of two kinds of force: the one physical, the other mental. The physical belongs to matter and is due to the properties with which it has been endowed; the other is the everywhere-present and ever-acting mind of God. To the latter are to be referred all the manifestations of design in nature and the ordering of events in Providence. This doctrine does not ignore the efficiency of second causes; it simply asserts that God overrules and controls them. Thus the Psalmist says: "I am fearfully and wonderfully made.... My substance was not hid from thee, when I was made in secret, and curiously wrought (or embroidered) in the lower parts of the earth."... "God makes the grass to grow, and herbs for the children of men." He sends rain, frost, and snow. He controls the winds and the waves. He determines the casting of the lot, the flight of an arrow, and the falling of a sparrow. [pp. 86–87]

Far be it from us to object to this mode of conceiving divine causation, although, like the two other theistic conceptions referred to, it has its difficulties, and perhaps the difficulties of both. But, if we understand it, it draws an unusually hard and fast line between causation in organic and inorganic nature, seems to look for no manifestation of design in the latter except as "God overrules and controls" second causes, and, finally, refers to this overruling and controlling (rather than to a normal action through endowment) all embryonic development, the growth of vegetables, and the like. He even adds, without break or distinction, the sending of rain, frost, and snow, the flight of an arrow, and the falling of a sparrow. Somehow we must have misconceived the bearing of the statement, but so it stands as one of "the three ways," and the right way, of "accounting for contrivances in nature," the other two being (1) their reference to the blind operation of natural causes, and (2) that they were foreseen and purposed by God, who endowed matter with forces which he foresaw and intended should produce such results, but never *interferes* to guide their operation.

In animadverting upon this latter view, Dr. Hodge brings forward an argument against evolution, with the examination of which our remarks must close:

> Paley indeed says that if the construction of a watch be an undeniable evidence of design, it would be a still more wonderful manifestation of skill if a watch could be made to produce other watches and, it may be added, not only other watches, but all kinds of time-pieces in endless variety. So it has been asked, if man can make a telescope, why cannot God make a telescope which produces others like itself? This is simply asking whether matter can be made to do the work of mind. The idea involves a contradiction. For a telescope to make a telescope supposes it to select copper and zinc in due proportions and fuse them into brass, to fashion that brass into inter-entering tubes, to collect and combine the requisite materials for the different kinds of glass needed, to melt them, grind, fashion, and polish them,[to] adjust their densities and focal distances, etc., etc. A man who can believe that brass can do all this might as well believe in God. [pp. 87–88]

If Dr. Hodge's meaning is that matter unconstructed cannot do the work of mind, he misses the point altogether; for original construction by an intelligent mind is given in the premises. If he means that the machine cannot originate the power that operates it, this is conceded by all except believers in perpetual motion, and it equally misses the point; for the operating power is given in the case of the watch, and implied in that of the reproducing telescope. But if he means that matter cannot be made to do the work of mind in constructions, machines, or organisms, he is surely wrong. "*Solvitur ambulando,*" *vel scribendo;* he confuted his argument in the act of writing the sentence. That is just what machines and organisms are for, and a consistent Christian theist should maintain that is what all matter is for. Finally, if, as we freely suppose it, he means none of these, he must mean (unless we are much mistaken) that organisms originated by the Almighty Creator could not be endowed with the power of producing similar organisms, or slightly dissimilar organisms, without successive interventions. Then he begs the very question in dispute, and that, too, in the face of the primal command, "Be fruitful and multiply," and its consequences in every natural birth. If the actual facts could be ignored, how nicely the parallel would run! "The idea involves a contradiction." For an animal to make an animal, or a plant to make a plant, supposes it to select carbon, hydrogen, oxygen, and nitrogen, to combine these into cellulose and protoplasm, to join with these some phosphorus, lime, etc., to build them into structures and usefully adjusted organs. A man who can believe that plants and animals can do this (not, indeed, in the crude way suggested, but in the appointed way) "might as well believe in God." Yes, verily, and so he probably will (in spite of all that atheistical philosophers have to offer) if not harassed and confused by such arguments and statements as these.

There is a long line of gradually-increasing divergence from the ultra-orthodox view of Dr. Hodge through those of such men as Sir William Thomson, Herschel, Argyll, Owen, Mivart, Wallace, and Darwin, down to those of Strauss, Vogt, and Büchner. To strike the line with telling power and good effect, it is necessary to aim at the right place. Excellent as the present volume is in motive and tone, and clearly as it shows that Darwinism may bear an atheistic as well as a theistic interpretation, we fear that it will not contribute much to the reconcilement of science and religion. . . .

Bibliography

This select bibliography is divided into three parts: (1) works with substantial treatment of Hodge, Princeton, and evolution; (2) twentieth-century commentary on Hodge's *What Is Darwinism?* and (3) works on Darwin, Darwinism, and evolution.

1. Works with Substantial Treatment of Hodge, Princeton, and Evolution

The following contain the most important discussions of Charles Hodge's *What Is Darwinism?* as well as of the immediate context in which that work was done.

Gundlach, Bradley J. "The Evolution Question at Princeton, 1845–1888." M.A. thesis, Trinity Evangelical Divinity School, 1989.

Illick, Joseph E., III. "The Reception of Darwinism at the Theological Seminary and the College at Princeton, New Jersey." *Journal of the Presbyterian Historical Society* 38 (1960): 152–65, 234–43.

Johnson, Deryl F. "The Attitudes of the Princeton Theologians toward Darwinism and Evolution from 1859 to 1929." Ph.D. diss., University of Iowa, 1969.

Livingstone, David N. "B. B. Warfield, the Theory of Evolution and Early Fundamentalism." *Evangelical Quarterly* 58 (Jan. 1986): 69–83.

———. "Darwinism and Calvinism: The Belfast-Princeton Connection." *Isis* 83 (1992): 408–28.

———. *Darwin's Forgotten Defenders: The Encounter between Evangelical Theology and Evolutionary Thought.* Grand Rapids: Eerdmans; and Edinburgh: Scottish Academic Press, 1987.

———. "Evangelicals and the Darwinian Controversies: A Bibliographical Introduction." *Evangelical Studies Bulletin* 4 (Nov. 1987): 1–6.

———. The Idea of Design: The Vicissitudes of a Key Concept in the Princeton Response to Darwin." *Scottish Journal of Theology* 37 (1984): 329–57.

Noll, Mark A., ed. *The Princeton Theology, 1812–1921: Scripture, Science, and Theological Method from Archibald Alexander to Benjamin Warfield*. Grand Rapids: Baker, 1983.

Scott, William Berryman [grandson of Charles Hodge]. *Some Memories of a Palaeontologist*. Princeton: Princeton University Press, 1939.

Smith, Gary S. "Calvinists and Evolution, 1870–1920." *Journal of Presbyterian History* 61 (Fall 1983): 335–52.

Wells, Jonathan. "Charles Hodge on the Bible and Science." *American Presbyterians* 66 (1988): 157–65.

———. *Charles Hodge's Critique of Darwinism: An Historical-Critical Analysis of Concepts Basic to the 19th Century Debate*. Lewiston, N.Y.: Edwin Mellen, 1988.

2. Twentieth-Century Commentary on Hodge's *What Is Darwinism?*

Briefer discussions of *What Is Darwinism?* are contained in the following works, many of which also treat larger theological or historical questions related to evolution and Christian faith.

Barbour, Ian G. *Issues in Science and Religion*. Englewood Cliffs, N.J.: Prentice-Hall, 1966.

Boller, Paul F., Jr. *American Thought in Transition: The Impact of Evolutionary Naturalism, 1865–1900*. Chicago: Rand McNally, 1969.

Brooke, John H. *Science and Religion: Some Historical Perspectives*. Cambridge: Cambridge University Press, 1991.

Conkin, Paul K. *American Christianity in Crisis*. Waco, Tex.: Baylor University Press, 1981.

Dillenberger, John. *Protestant Thought and Natural Science: A Historical Interpretation*. Garden City, N.Y.: Doubleday, 1960.

Dupree, A. Hunter. *Asa Gray, 1810–1888*. Cambridge, Mass.: Harvard University Press, 1959.

Foster, Frank H. *The Modern Movement in American Theology: Sketches in the History of American Protestant Thought from the Civil War to the World War*. New York: Revell, 1939.

Gillespie, Neal C. *Charles Darwin and the Problem of Creation*. Chicago: University of Chicago Press, 1979.

Greene, John C. "Darwinism as a World View." In idem, *Science, Ideology, and World View: Essays in the History of Evolutionary Ideas*, 128–57. Berkeley: University of California Press, 1981.

Gregory, Frederick. "The Impact of Darwinian Evolution on Protestant Theology in the Nineteenth Century." In *God and Nature: Historical Essays on the Encounter between Christianity and Science*, edited by David C. Lindberg and Ronald L. Numbers, 369–90. Berkeley: University of California Press, 1986.

Hoeveler, J. David, Jr. *James McCosh and the Scottish Intellectual Tradition.* Princeton: Princeton University Press, 1981.
Hofstadter, Richard. *Social Darwinism in American Thought.* Rev. ed. Boston: Beacon, 1955.
Hovenkamp, Herbert. *Science and Religion in America, 1800–1860.* Philadelphia: University of Pennsylvania Press, 1978.
Lindberg, David C., and Ronald L. Numbers. "Introduction." In *God and Nature: Historical Essays on the Encounter between Christianity and Science,* edited by David C. Lindberg and Ronald L. Numbers, 1–18. Berkeley: University of California Press, 1986.
Moore, James R. *The Post-Darwinian Controversies: A Study of the Protestant Struggle to Come to Terms with Darwin in Great Britain and America, 1870–1900.* Cambridge: Cambridge University Press, 1979.
Phipps, William E. "Asa Gray's Theology of Nature." *American Presbyterians* 66 (Fall 1988): 167–75.
Roberts, Jon H. *Darwinism and the Divine in America: Protestant Intellectuals and Organic Evolution, 1859–1900.* Madison: University of Wisconsin Press, 1988.
Russett, Cynthia E. *Darwin in America: The Intellectual Response, 1865–1912.* San Francisco: W.H. Freeman, 1976.
Wilkins, Walter J. *Science and Religious Thought: A Darwinian Case Study.* Ann Arbor: UMI Research Press, 1987.
Wilson, R. Jackson. *Darwinism and the American Intellectual: An Anthology.* 2d ed. Belmont, Calif.: Wadsworth, 1989.

3. Works on Darwin, Darwinism, and Evolution

The following works make up an introduction to the immense bibliography that now exists on the subject of Darwinism. They are pertinent in specific theological or historical ways to Charles Hodge's study of Darwinism or to issues discussed in the Introduction.

Bowler, Peter J. *Charles Darwin: The Man and His Influence.* Cambridge, Mass.: Blackwell, 1990.
———. *Evolution: The History of an Idea.* Berkeley: University of California Press, 1984.
———. *The Non-Darwinian Revolution.* Baltimore: Johns Hopkins University Press, 1988.
Bozeman, Theodore D. *Protestants in an Age of Science: The Baconian Ideal and Antebellum American Religious Thought.* Chapel Hill: University of North Carolina Press, 1977.
Brooke, John H. "The Relations between Darwin's Science and His Religion." In *Darwinism and Divinity,* edited by John Durant, 40–75. New York: Basil Blackwell, 1985.

Darwin, Charles. *The Origin of Species: A Variorum Text.* Edited by Morse Peckham. Philadelphia: University of Pennsylvania Press, 1959.

Desmond, Adrian, and James R. Moore. *Darwin.* New York: Warner, 1991.

Hull, David L. *Science as a Process: An Evolutionary Account of the Social and Conceptual Development of Science.* Chicago: University of Chicago Press, 1988.

Johnson, Phillip E. *Darwin on Trial.* Washington, D.C.: Regnery Gateway; and Downers Grove, Ill.: Inter-Varsity, 1991.

Kohn, David, ed. *The Darwinian Heritage.* Princeton: Princeton University Press, 1985.

Livingstone, David N. "The Darwinian Diffusion: Essay Review." *Christian Scholar's Review* 19 (Dec. 1989): 186–99.

———. "Evolution as Metaphor and Myth." *Christian Scholar's Review* 12 (1983): 111–25.

———. "Evolution, Eschatology and the Privatization of Providence." *Science and Christian Belief* 2 (1990): 117–30.

———. *The Preadamite Theory and the Marriage of Science and Religion.* Transactions of the American Philosophical Society 82.3. Philadelphia: American Philosophical Society, 1992.

Moore, James R. "Darwin's Genesis and Revelations" [essay review on publications related to Darwin's correspondence]. *Isis* 76 (1985): 570–80.

———. "Deconstructing Darwinism: The Politics of Evolution in the 1860s." *Journal of the History of Biology* 24 (1991): 353–408.

———. "Speaking of 'Science and Religion'—Then and Now." *History of Science* 30 (1992): 311–23.

———, ed. *History, Humanity, and Evolution: Essays for John C. Greene.* Cambridge: Cambridge University Press, 1990.

Numbers, Ronald L. *The Creationists.* New York: Knopf, 1992.

———. "George Frederick Wright: From Christian Darwinist to Fundamentalist." *Isis* 79 (1988): 624–45.

Ospovat, Dov. "God and Natural Selection: The Darwinian Idea of Design." *Journal of the History of Biology* 13 (1980): 169–94.

Plantinga, Alvin, et al. "Creation/Evolution and Faith." *Christian Scholar's Review* 21 (Sept. 1991): 3–114.

Richards, Robert J. *The Meaning of Evolution: The Morphological Construction and Ideological Reconstruction of Darwin's Theory.* Chicago: University of Chicago Press, 1992.

Young, Davis. "Scripture in the Hands of Geologists." *Westminster Theological Journal* 49 (1987): 1–34, 257–304.

Young, Robert M. *Darwin's Metaphor: Nature's Place in Victorian Culture.* Cambridge: Cambridge University Press, 1985.

Index

Absent God, theory of, 75, 76, 87–88, 166
Accidental variations, 92, 94–95, 163–64
Agassiz, Louis, 21, 49, 51, 52–54, 113, 117–19, 131, 133, 143, 146, 148, 155
Alexander, Archibald, 17, 19
American Church Review, 32
America toward Darwinism, shifting attitudes in, 25
Amoebas, 81
Animals: gulf between plants and, 139; sudden appearance of new kinds of, 148
Ant, slave-making, 82
Ape as the progenitor of man, 50, 84
Argyll, Duke of, 76–77, 99, 105n, 115–17, 166
Arnold, Matthew, 89
Artificially produced varieties to simple form, reversion of, 147
Artificial selection, 85, 86
Ascending order in creation, 145–46
Ateleological, interpretation of Darwinism as, 36–38, 39–40
Ateleology, 29. *See also* Design in nature, question of
Atheism: Darwinism as, 12, 29, 30n, 34, 38, 44, 46, 56, 109, 114, 118, 155, 156–57, 160–66, 169; Huxley's reply to charge of, 105n

Bacon, Francis, 12n, 14
Baconian scientific induction, 14, 45, 55n, 56
Baer, K. E. Von, 154
Baptist Quarterly, 32
Barbour, Ian G., 39n
Beale, Lionel, 132
Beauty: Darwin's explanation of, 98; as difficulty for natural selection, 99, 100; as evidence of design, 116
Bee, cell-building, 82
Berkeley, George, 75n
Berry, R. J., 47n
Bible and science. *See* Scripture and science
Biblical Repertory and Princeton Review. See *Princeton Review*
Bischoff, Theodor, 90
Boller, Paul F., Jr., 36n
Boscovich, Ruggiero, 70, 71n
Bowen, Francis, 37, 38
Bowler, Peter J., 40n
Brahe, Tycho, 55n
Bree, C. R., 134, 146n
Bridgewater Treatises, 44
British Association for the Advancement of Science, 37
Brooke, John H., 39n

175

Brougham, Henry, 87
Brown, J. C., 26–27
Büchner, Ludwig, 68, 108, 109, 119–20, 140, 155
Buddhism, 142
Burnett, Alexander, 27
Butler, Bishop, 162

Cabell, J. L., 20
Calvinist convictions, 13–14, 18, 22; and Darwinism, 46n, 160
Cambrian period, 79n, 80
Carpenter, William B., 135
Catastrophism, 51n, 144n
Causation in organic and inorganic nature, 167
Causes: God and physical, 65; in Darwinism, efficiency of physical, 78, 160
Chalmers, Thomas, 57
Chambers, Robert, 90n
Christian Darwinism, 34
Christian evolutionist, 34–35
Christian Scholar in the Age of the Reformation (Harbison), 13
Clark, Joseph, 51–52
Classification, diversity in, 143
Commonsense philosophy, 14, 21, 45, 64n, 75n, 103n
Concursus, 77n
Conkin, Paul K., 39n
Consciousness: intuitions of human, 14, 21, 31–32, 45, 64n, 132, 137, 153; in Spencer, 71–72; origin of, 149, 151–52
Contrivance: methods of accounting for, 86–89, 100, 162, 167; as difficulty for natural selection, 99, 154; as evidence of design, 115
Creation: Mosaic account of, 27, 57, 59, 146, 148n; of substance and creation of form, distinction between (Haeckel), 111; choice between evolution and, 128
"Creationists," modern, 30n
Creator, Darwin's admission of, 36, 78, 93, 150, 152, 155, 160, 162
Croll, James, 79
Cuvier, Georges, 51, 121

Dana, James Dwight, 57, 59, 144
Darwinism: shifting attitudes in America toward, 25; definition of, 28, 34–47, 160–61; as hypothesis (Dawson), 28; as teleological, 29n, 33, 34, 36, 40n, 46n, 94n, 159; distinguished from evolution, 34, 91, 119, 165; skepticism regarding "essential," 39–42; and Calvinism, 46n, 160; elements of, 89–92, 138; incredibility of, 96, 139–40; as hypothesis (Huxley), 103–5, 144–45, 147; to religion, relation of, 129–57; objections to, 138–57; unprovability of, 140; denial of design as reason for success of, 140–43
Darwin's response to *What Is Darwinism?* 33
Dawson, John William, 27–28, 46, 126–29, 145, 147, 148, 149, 151
Day-age theory, 57
Definition of Darwinism, 28, 34–47, 160–61
Derivation, 142–43
Descent of Man (Darwin), 83–84
Design in nature: question of, 29–30, 34, 36–37, 85, 86, 92, 95, 98, 99, 102, 112–14, 115, 124, 125, 126, 138, 163, 165, 167; Hodge on, 42–46; argument *to,* 44; in evolutionary theory, 47n; denial of, as reason for Darwinism's success, 140–43; denial of, as grand objection to Darwinism, 152–56; Darwin's admission of, 153. *See*

also Teleology
Deus ex machina, 75
Developmental hypothesis (discussion by the Evangelical Alliance), 26–27
Dillenberger, John, 36n
Diversity in classification, 143
Dobzhansky, Theodosius, 47n
Dualistic theory of the universe, 112–13; Darwin's admission of, 152
Dubois-Reymond, Emil, 151
Dupree, A. Hunter, 36n

Earth, age of the, 25, 57
Edwards, Jonathan, 22, 75n
Eldredge, Niles, 41
Engis, 147
Epicurean theory, 68–69
Epicurus, 108
Equilibrium, punctuated, 40
"Essay on Classification" (Agassiz), 113
Essence of Darwinism, 38, 39–42
Evangelical Alliance, 23–28
Everything in relation to the world, theory that God does, 75–77, 166
Evidence, meaning of, 130–32
Evolution: as compatible with religion, 26; distinguished from Darwinism, 34, 91, 119, 165; as compatible with Scripture, 35; design in, 47n; as element of Darwinism, 89–91; progressive, 90n; as compatible with design but not with Scripture, 125, 138; choice between creation and, 128; and theism, 128, 163, 164; half-way, 128; gaps in theory of, 149–52
"Examination of Some Reasonings against the Unity of Humankind" (Hodge), 21n, 49
Explanations (theories), distinction between facts and, 58, 59, 133–34, 137
Eye, formation of the, 28, 45, 92, 95–96, 106, 108, 109, 125

Facts: Hodge's emphasis on scientific, 31–32; Bible's accordance with, 55; distinction between theories and, 58, 59, 133–34, 137; of nature, restriction of science to, 130–31; of geological record, 148–49
Faith and science, 13, 111–12, 131
Fertilization of plants, 97–98
Feuerbach, Ludwig, 68
Final cause, denial of, 30, 92, 98, 113, 155. *See also* Design; Teleology
Fish, ancient, 146, 148
Fixedness of species, 46, 51, 113n, 134, 143–49, 156
Flight of birds, contrivance in the, 117
Flourens, M., 121–22
Force (in Spencer), 69, 72, 74, 89
Foster, Frank H., 36n

Galileo, 55, 133
Gaps in the theory of evolution, 149–52
Gap theory, 57
Geocentricism, 55n, 59
Geoheliocentricism, 55n
Geological ages, 145, 146n
Geological record, facts of, 148–49
Geology and the Bible, 57, 58–59
Germ, primordial, 51, 78, 139, 152, 161
Gillespie, Neal, 36, 37, 38
Gish, Duane, 30n
Gliddon, George, 49, 50
God: the word, 64; theory of absent, 75, 76, 87–88, 166; does nothing in relation to the world, theory

that, 75, 76; does everything in relation to the world, theory that, 75–77, 166
Gott und die Natur (Ulrici), 90
Gould, Stephen Jay, 41
Gradual species transformation, Darwinism as, 40
Graham, William, 19
Gray, Asa, 24n, 29n, 32, 33, 34–35, 36, 40, 41, 46n, 94, 101n, 155–56; response to *What Is Darwinism?* 159–69
Green, Ashbel, 17, 19
Green, William Henry, 146n
Greene, John C., 39n
Gregory, Frederick, 36n
Guyot, Arnold, 27, 57

Haeckel, Ernst, 108, 110–14, 131, 136, 140, 143, 153, 155
Hamilton, William, 72–73, 153n
Harbison, E. Harris, 13
Harvey, William, 124
Heliocentric system, 55n, 59
Helmholtz, Hermann von, 141
Henry, Joseph, 18, 19–20, 102
Henslow, George, 40, 90–91, 124–25, 131, 138
Heredity, law of, 79, 99, 153–54, 161
Herschel, John, 76, 166
Higher teleology, 108, 109
Hodge, Archibald Alexander, 16n, 29, 42, 67n
Hodge, Charles: life of, 16–17; writings of, 17–18; interest in science of, 18–23, 30, 57
Hodge, Hugh Lenox, 16–17, 19
Hodge, Mary Blanchard, 16–17
Hoeveler, J. David, Jr., 35n
Hoffmann, Friedrich, 90
Hofstadter, Richard, 35n
Holbach, Baron Paul d', 89
"Homology," 50

Horse, foot of the, 146
Hovenkamp, Herbert, 35n
Hull, David, 41
Humanity: unity of, 20, 49–51, 54n, 55, 56; multiple origins for, 21, 49–51, 54n, 55, 113n, 132. *See also* Man
Humboldt, Alexander von, 52–54
Hume, David, 14, 71, 142
Humming-birds, 116
Huxley, Thomas Henry, 40, 63, 90, 98, 102–9, 114, 122, 135, 136, 138, 144, 147, 150, 153
Hybrids, 147
Hylozoic theory, 74–75
Hypothesis: Dawson's categorization of Darwinism as, 28; Huxley's categorization of Darwinism as, 103–5, 144–45, 147; distinction between theories and, 103

Immutability of species. *See* Fixedness of species
Inerrancy of the Bible, 53
Infallibility of Scripture, 58, 107n
Inspiration, biblical, 52, 53, 58
Instincts, origin of animal, 81–82
Intelligence: evolution of, 84; implied by design, 152–53
Interpretations of the Bible, modification in, 21, 31, 52, 54–55, 58, 59
Intervention in nature, question of divine, 33, 56, 110, 120, 128, 150, 159, 165, 166
Intuitions, 14, 21, 31–32, 45, 64n, 132, 137, 153. *See also* Consciousness

Jacobi, Friedrich, 73
Janet, Paul, 119–21
Johnson, Phillip, 40

Kelvin, Lord. *See* Thomson, William

178

Index

Knowledge, natural and revealed, 12–13, 54. *See also* Science and Scripture

Lamarckianism, 23, 24n, 69n, 90, 108n
Laplace, Pierre de, 101
Laws, Darwinian, 79, 93, 99–100; established by divine power, 162
Le Conte, Joseph, 66n
Leibniz, Gottfried von, 70, 71n
Lewontin, Richard C., 40
Life: in Darwinism, existence of, 78, 161; struggle for, 79, 99, 129, 161; Huxley's definition of, 103n; origin of (problem for science), 149–51
Life-in-matter theory, 74–75
Lindberg, David, 38
Links, no evidence of, 148, 149. *See also* Transition state
Literalist, Hodge viewed as, 35
Livermore, L. J., 32
Lowest forms of life, 81
Lyell, Charles, 40, 133–34, 150n

McCosh, James, 23–24, 26, 29, 35, 101n
MacKay, Donald, 47n
Malebranche, Nicolas de, 75n
Man: in the scriptural theory of the universe, 65, 66; in pantheism, 67–68; origin of, 83–85, 110, 141; as exception to doctrine of evolution, 102, 106–7; evolution's failure to account for the higher lineaments of, 128–29; antiquity of, 146–47. *See also* Humanity
Mansel, Henry, 71–73
Materialism, 89, 112, 114, 119; Huxley and, 102n
Matter in Darwinism, existence of, 78

Matthew, Patrick, 91, 92n
Mechanical view of the universe, 112–14
Meteors as origin of life, 151
Methodist Quarterly Review, 32
Mill, John Stuart, 153
Miller, Hugh, 57, 148
Mind, evidence of constant activity of, 87
Miracles, 77; Huxley on, 135; Strauss on abolition of, 140–41
Misery as difficulty for theism, world, 66n
Mitchell, Walter, 45, 122–26, 142
Mivart, St. George, 106–7
Moleschott, Jacob, 68
Mollusks, 145
Monistic theory of the universe, 112–14, 136
Moore, James, 39, 46
Moralists, 110
Moral materialism, 112
Moral nature, evolution of, 84
Morris, Henry M., 30n
Morton, Samuel George, 49, 51, 144
Mosaic account of creation, 27, 57, 59, 146, 148n
Müller, Johannes, 90
Multiple origins for humanity, 21, 49–51, 54n, 55, 113n, 132

Natural, word, 30, 85–89, 162
Natural and revealed knowledge, 12–13, 54. *See also* Science and Scripture
Natural causes, blind operation of, 89, 98, 141, 152, 163, 167
Natural History of Creation (Haeckel), 108, 111, 136n
Natural selection, 28, 29, 30, 37, 79–89, 91–92, 95–96, 107, 120, 142, 161, 164, 165; Darwinism as evolution exclusively by, 41; objections

179

to (Argyll), 99; possibility of direction in, 120
Natural Selection Not Inconsistent with Natural Theology (Gray), 41
Natural theology, 87n; Hodge and, 43–44
Natural Theology (Paley), 87n, 153
Nature, Darwin's personification of, 96n, 115, 121–22
Nebular Hypothesis, 25, 101
Neo-Platonism, 66, 67n
New Jersey, College of, 17, 23, 130n
New kinds of animals, sudden appearance of, 148
Newton, Isaac, 14, 70, 71n
New York Observer, 21, 51–56
Nothing in relation to the world, theory that God does, 75, 76
Nott, Josiah, 49, 50
Numbers, Ronald, 38

Old Faith and the New (Strauss), 140
On the Classification of the Mammalia (Owen), 50
Orchids, 97, 100
Order, law of, 66n
Organs, undeveloped or useless, 66n, 95
Origin of Species, Hodge's first comment on, 49–51
Over-Production, law of, 79, 99, 161
Owen, Richard, 50, 90, 142–43

Paine, Thomas, 22
Paley, William, 44, 45, 87, 88n, 153, 168
Pantheistic theory, 66–68
Peabody, Andrew, 118–19, 155
Perfection: of every species, 145–46; of earliest remains of man, 147
Personification of nature, Darwin's, 96n, 115, 121–22
Phipps, William E., 35n

Physical causes: God and, 65; in Darwinism, efficiency of, 78, 160
Pictet de la Rive, François, 145
Pigeons, 85
Plant and animal, gulf between, 139
Pollenization, 97
Polygenism, 21, 49, 55
Porter, Noah, 111
Primordial form, 144
Primordial germ, 51, 78, 139, 152, 161
Princeton approach, 12–16, 22
Princeton Review, 17, 20, 32, 49–53, 62
Princeton Theological Seminary, 11–12, 17, 56
Princeton University. *See* New Jersey, College of
Principles of Geology (Lyell), 133n, 134
Progressive evolution, 90n
Protoplasm, 102–3n, 150
Protozoa, 81
Ptolemy, 55n, 58
Public life and theology, 15, 22–23, 25
Punctuated equilibrium, 40
Purpose in nature. *See* Design in nature, question of

Reason, origin of, 149, 151
Reddie, James, 123n
Reformed convictions, 13–14, 18. *See also* Calvinist convictions
Reid, Thomas, 14
Reign of Law (Duke of Argyll), 99, 105n
Religion: as compatible with evolution, 26; relation of Darwinism to, 129–57; causes of the alienation between science and, 129–38; as both knowledge and feeling, 138
Revealed and natural knowledge,

12–13, 54. *See also* Science and Scripture
Richards, Robert, 40
Roberts, Jon, 25
Russett, Cynthia E., 36n

Saltationism, 40
Schmerling, P. C., 147n
Science: and faith, 13, 111–12, 131; and Scripture, 15, 20–21, 31, 47n, 51–59, 122n, 123; Hodge's interest in, 18–23, 30, 57; causes of the alienation between religion and, 129–38; meaning of, 130–32; stumbling-blocks to, 149–52. *See also* Natural and revealed knowledge
Sciences, unity of, 130n, 137
Scientific men toward religion, hostility of, 134–38
Scott, William Berryman, 159
Scriptural method of accounting for contrivance, 86
Scriptural theory of the universe, 64–66, 123–24
Scripture: and science, 15, 20–21, 31, 47n, 51–59, 122n, 123; infallibility of, 58, 107n. *See also* Revealed and natural knowledge
Sebright, John, 107
Secondary causes, 75, 77, 86, 163, 166, 167
Senses, restriction of science to facts perceived by the, 131–32, 137
Sin, 66n
"Skepticism of Science" (Clark), 51
Smith, Samuel Stanhope, 55n
Society and theology, 15, 22, 25
Spanner, Douglas, 47n
Species: fixedness of, 46, 51, 113n, 134, 143–49, 156; definition of, 143; criteria of, 144
Spencer, Herbert, 69–74, 78, 80, 89, 99, 128, 135, 142
Spinoza, Benedict de, 67, 141
Spontaneous generation, 78, 139, 150
Sterile females in insect communities for natural selection, difficulty of, 82–83
Stewart, Dugald, 14
Stirling, James H., 150
Strauss, David, 68, 140–42, 149, 151–52, 155, 157
Struggle for life, 79, 99, 129, 161
Supernatural selection, 85
Survival of the fittest, 79, 80, 95, 99, 129, 142, 161. *See also* Natural selection
Systematic Theology (Hodge), 18, 53, 56–59

Teilhard de Chardin, Pierre, 47n
Teleological: Darwinism as, 29n, 33, 34, 36, 40n, 46n, 94n, 159; language, Darwin's use of, 86, 115
Teleology, 29; in Hodge's theology, 42–46; rejection of, as distinctive element of Darwinism, 89–129; higher, 108, 109. *See also* Design in nature, question of
Theism: unscriptural, on the nature of the universe, 75–77; and evolution, 128, 163, 164
Theology and public life, 15, 22–23, 25
Theories: distinction between facts and, 58, 59, 133–34, 137; regarding the universe, 64–89, 112–14, 123–24; distinction between hypotheses and, 103
Thompson, Henry, 135
Thomson, William, 150–51, 153
Transactions of the Victoria Institute, 123
Transcendental, matter as, 74

Transformation, Darwinism as gradual species, 40
Transition state, no evidence of, 145, 149. *See also* Links, no evidence of
Transmutation of species, evidence against, 143–49
Trilobites, 145
Tyndall, John, 37, 74, 135
Types of Mankind (Nott and Gliddon), 50

Ulrici, Hermann, 90, 131
Undeveloped organs, 66n, 95
Uniformitarianism, 144n, 150n
Unintentional variations, 92, 94–95, 163–64
Unitarian Review and Religious Magazine, 32
Unity: of humanity, 20, 49–51, 54n, 55, 56; of the sciences, 130n, 137
"Unity of the Church" (Hodge), 25
Universe: theories about the, 64–89, 112–14, 123–24; origin of the, 65, 69–70, 101; self-evolution of the, 108
Unscriptural theism on the nature of the universe, 75–77
Unused organs, 66n, 95

Van Till, Howard, 47n
Variation, law of, 79, 99, 161
Variations: accidental, 92, 94–95, 163–64; that are injurious before full development, 117

Varieties artificially produced to simple form, reversion of, 147
Vestiges of Creation (Chambers), 90, 125, 140, 148
Victoria Institute, 122–23
Vitality, 103n, 113
Vogt, Karl, 68, 69n, 83–84, 108, 109–10, 135, 140, 155
Voyage of a Naturalist (Darwin), 31

Wagner, Rudolf, 90
Wallace, Alfred Russel, 40, 76, 90, 98–102, 106–7, 146
Warfield, Benjamin, 11–15, 29, 42, 77n, 87n, 95n, 101n, 146n
Warington, George, 123n, 126
Watts, Robert, 37–38
Way of Life (Hodge), 18
Weis, Ludwig, 142
Weldon, George W., 26
Wells, Jonathan, 43–44
Wells, W. C., 91
What Is Darwinism? background to writing of, 23–28; basic concerns of, 29–32; reception of, 32–33; present-day reaction to, 35–39
Whewell, William, 87, 124, 162
Whitcomb, John C., 30n
Wilkins, Walter, 36–37, 38
Wilson, R. Jackson, 35n
Wissenschaft, 130
Witherspoon, John, 14, 17, 19
Wright, George Frederick, 46n, 101n, 160
Wunder, 140